Alternative Light Styles

Single, duplex, and round-wick kerosene lamps.

Alternative Light Styles

The Lamp Keeper's Guide to Illumination Beyond the Power Lines

Tim Matson

Photographs by the author

THE COUNTRYMAN PRESS
Woodstock, Vermont

Copyright © 1984 by Tim Matson

All rights reserved. No part of this book may be reproduced in any form or by any means—electronic, mechanical, photocopying, or other—without permission in writing from the publisher. Nor may any part of this book be stored in a retrieval system or transmitted in any form or by any means without permission in writing from the publisher, except for reviewers, who may quote brief passages or reproduce illustrations in an article or critical review.

Parts of *Alternative Light Styles* have appeared in *Harrowsmith, Fine Homebuilding, Country Journal,* and *Yankee.*

Cover: Shepherd's candle lantern by Anne Margolis

Library of Congress Cataloging in Publication Data
Matson, Tim, 1943–
   Alternative light styles.

   1. Gas-lighting.   2. Candles.   I. Title.
TH7953.M37   1984   621.32'3   84-19909
ISBN 0-88150-029-1 (pbk.)

Editors: Peter Jennison, Pat Peat
Design & production: David Robinson
Photo finishing by Sun Photo Lab, Hanover, New Hampshire
Composed by Ampersand Typesetters, Inc.
Printed in the United States by Capital City Press

Contents

An Explanation	1
Candlepower	5
The Kerosene Wick Lamp	11
Aladdin Light	22
Lantern Light	37
Liquid Propane Gaslight	49
Incandescent Mantles	61
The Return of Firelight	73
Sources	81
Index	93

This Book is for my Mom and Dad

An Explanation

Since May 1974, I haven't switched on an electric light. I cut my electric habit with a vengeance: cold turkey when I was lucky, but mostly pork and beans warmed over a campfire or a candle. By day I built my cabin in the light of the long North Country summer sun, working with a chain saw and hand tools. At dusk I kindled a fire near the tent. The orange flames threw enough illumination for me to scratch out a few letters and to sharpen the chain saw before the fire died. Sometimes rain sizzled the fire, and I burrowed into the tent. A burning taper cradled in a clay pot beamed spokes of light through holes in the side of the pot. The tent sparkled.

By Halloween I had the cabin closed up. For a brighter night life I brought in a trio of kerosene lamps, which illuminated the early years. Eventually I replaced some of the lamps with brighter liquid propane (LP) mantle lights. (Without a doubt, if you plan to light by fire for primary illumination, LP is tops.) Thus I passed through a condensed evolution of lighting, progressing from Neanderthal campfires to twentieth-century mantle lamps. Naturally I considered the next step. That's when I discovered it would cost me close to $3,000 simply to run the electric line up my hill. Then come the bills, the inevitable appliances, the power failures. No, the power line wouldn't do.

I am not alone in this decision. Scattered throughout this rural town of 672 souls are close to a score of families living year-round without commercial, power company electricity; most light primarily with kerosene lamps and LP, a few with photovoltaic and windmill power. The locally generated electricity is experimental and expensive. Firelight is inexpensive, reliable, and mobile.

Besides cutting the utility bill, those of us outside the main current often find that "non-electric land" is a very good buy. The soil is likely to be richer and the forest thicker. Certainly the price tag will be lower, along with the taxes, not to mention fuel bills. It costs me less than $100 a year to power the lights. But don't get me wrong. With gas-powered lights, stove, and refrigerator, and fire view wood stoves, I can't exactly call myself a primitive. An energy vegetarian, perhaps.

And this is just the tip of the iceberg. Electric rates nationwide are soaring. An estimated 25 million mantle lamps burn in this country annually, excluding kerosene wick lamps and lanterns. There are year-round lamp keepers, and people who keep firelights for back up use during power breaks, weekends or vacations at a country home, and camping, boating, and special occasions. And special occasions may not always be a festive dinner by candlelight. In the summer of 1982, for instance, 30,000 people in Connecticut were cut off for not paying their electric bills.

Nowadays the Coleman Company makes more than a million lanterns annually. Aladdin is going strong after three-quarters of a century. LP gaslights, which until recently turned up mostly in recreational vehicles and trailers, are now moving into homes and vacation camps. Humphrey Products of Kalamazoo, Michigan, has manufactured LP gaslights since 1901, when they supplied much of the outdoor lighting in the United States. I asked Herb Rosenhagen, president of Humphrey, to give me an estimate of the Humphrey lights now in operation.

"It would have to be high in the hundreds of thousands," he said. "That's in the United States, Canada, South America, and the Bahamas. You have to remember that these lights last thirty years. Last year our sales were up 42 percent over the year before."

During the development of my firelight system, it occurred to me that it would have been made immensely easier—and safer—with a guidebook. Too often the lamp keeper's craft is won at the sacrifice of broken chimneys, cracked mantles, or kerosene stains. Worse, an improperly installed or unmonitored lamp can kill you. I had searched for a book about firelight but found only history that stopped short at about 1900 with the inception of electricity. Electric light seems to have blinded many people to the reality that firelight continues to illuminate homes around the world. Three-quarters of the earth's people still light by fire.

It's interesting also to recall that the Welsbach mantle appeared almost simultaneously with Edison's electric bulb. The Welsbach mantle quadrupled the candlepower of the wick lamps and the gaslights of the time and outperformed electric light during the early century. In rural areas, gaslights, Aladdins and Colemans continued to rule the night as late as the fifties. In the history of artificial lighting, the reign of electricity is just the blink of an eye. Indeed, the coming of electrification often has been viewed with suspicion.

Clearly, a fresh look at non-electric light is needed. Advertisements and instruction pamphlets for most lamps highlight the positive side of the unit, of course, but often fail to mention the difficulties: the need for adequate ventilation, flare-ups, mantle protection and radiation, proper fueling, and so forth. Many suppliers have had little experience with their products. For instance, one distributor I know recommended lighting LPs from above, a flat-out dangerous procedure.

I also want to scotch the myth of electric efficiency. When I mentioned to physicist recently that I use gaslight, his fuse seemed to blow.

"Gaslight is ten times as wasteful as electric light!" He exploded. "Gaslight disappeared for a good reason. It's grossly inefficient!"

Not quite. Most of the electricity produced today is generated from coal, oil, or uranium—all non-renewable and polluting— and about two-thirds of the energy content in these fuels winds up as wasted heat, not electricity. If that's efficient, we need a new definition of the word. Instead, what we really need is a new source of energy. Photovoltaic cells and wind/hydro generators show promise, but they are costly. Perhaps the most promising "new" fuel is a gas—methane. Methane can be piped from deposits in the earth, or better yet, bacteriologically brewed as a form of solar fuel from manure, garbage, sewage, crop residues, and seaweed. It's easy to transport and can be burned directly or used to generate electricity.

"The fuel that most closely approaches the ideal is methane," wrote Barry Commoner in an essay in *The New Yorker* in May 1983.* It burns with the blue flame familiar from the kitchen stove or with the bright yellow flame of the gaslight that once illuminated streets, homes, and shops." Today the fuel burned in most North American wick and mantle lamps is kerosene, not methane. But as Commoner points out, hundreds of thousands of methane generators provide villages in Asia with fuel for lights, cooking, and some farmers in this country have adopted the technique. It would be simple to convert LP gaslights, for instance, to methane.

For now, firelight seems possessed of a schizophrenic nature. There's a classic aura about it that epitomizes the nineteenth-century pre-electric era; yet there's a post-industrial, twentieth century survivalist spirit, too. Look at the fuel burning lamp as a museum piece or as a pioneering element in your home. Either way, firelight burns like a torch passed through history, connecting the past—the time we first dared to kindle our own suns—and what's to come.

*("A Nearly Perfect Fuel," *The New Yorker*, May 2, 1983)

Candlepower

In the beginning—before chimney lamps or electricity—was the candle. And when the power fails, the candle reappears. It is the one truly independent light: fuel and wick fused into a single element, no parts to lose or break, easy to transport, and impossible to explode. No mistaking empty fuel tanks either; when a candle's done it disappears. Thus candles lead the pack in foolproof emergency lighting. A twelve-inch household paraffin candle will burn for seven or eight hours. Better yet are emergency candles designed for lengthy blackouts. Molded under sixteen hundred pounds of pressure, these stubby microcrystalline paraffin candles will burn fifty hours or more. Candlelight pales beside lamplight, of course. But it's simple enough to boost illumination by gathering several candlesticks in one place.

A Candle's Worth of Light

Like horsepower, the phrase "candlepower" hangs on in the language if not in science textbooks. Originally it reflected the power of pre-electric light. A candle's worth of light was deemed the fundamental unit of illumination: the sun at its peak produced 600,000 candlepower per square inch, a mantle light up to 100 candlepower, and kerosene lamp about 5 or 10 candlepower. But the measure suffered from problems of accuracy. A slender beeswax taper is no more kin to a fat tallow candle than a thoroughbred is to a Clydesdale. So a standard candle was established. Originally it was defined as the luminous intensity of several carbon filament lamps. In 1948 the current standard was adopted: 1/60th of the light intensity emitted by 1 square

Symbol of life, and the most reliable of all firelights, it's even designed to trim itself. As the candle burns, the braided wick twists out of the flame and the residue is incinerated. If some ash does settle on the wax, remove it. Otherwise, the candle will gutter and smoke.

centimeter of a so-called blackbody (or Planckian) radiator at the temperature at which platinum solidifies (2,046°K). It's called the new international candle, or *candela*, and has a nicer ring, I think, than *watt*.

Candlepower is good for more than blackouts. Scented with citronella or other insect repellant, a burning candle raises a defense against mosquitoes. A small candle under a kettle heats up a fondue or a stew pot. Sulfur candles are used to fumigate, and perfumed candles burn like incense. In the past, candles have beamed the light to write letters by, and then produced the seal. To light his easel at night, Michelangelo burned a candle in a holder worn on his head. Candles have been exchanged as currency and burned to measure time. Traditionally, eggs as well as wine were graded and inspected by candlelight. In ancient pagan rites and in modern churches, votive candles invoke divine power.

One of the original blessings of candlelight was the simplicity of manufacture. Candles could be molded at home with suet or tallow rendered from a homegrown beast—roughly 300 candles from an ox, for instance. Natural fat candles were valued doubly by early lighthouse keepers. If the food ran out during a storm they could eat the beacons. Today the connection between candlelight and food remains strong. More candles burn in restaurants than at home, and at home most are lit during dinner.

Beeswax

The finest candles of all are made from beeswax. A twelve-inch beeswax candle will burn for ten to twelve hours, several hours longer than a household paraffin candle of equal size. Beeswax candles, which are widely available, have less tendency to drip than paraffin, won't smoke, hold their shape better in hot weather, and yield a steadier flame. Best of all is the fragrance of beeswax, whether it's burning or not: a honey vanilla scent to match its golden color. Candles made from the capping wax that bees use to seal honeycombs burn the brightest and with the richest bouquet. I think the slightly higher cost of beeswax is more than justified by its longer, smokeless burn. And nothing beats beeswax candles for after-dinner marshmallow toasts.

Candle Cautions

There are several elemental precautions for the candle burner. It's important to steer away from excessive heat for candle placement and storage. In a sunny window, for instance, candles may bend. If not straightened, they'll be out of plumb and a nuisance, if not impossible, to burn. Immersion in a pan of warm water will make them pliable enough for straightening. Broken candles can be fixed by melting and resealing cracks after heating in water, or brief exposure to a flame. Wax that is hot enough to liquefy is hot enough to burn the skin, so caution is advised.

CANDLEMAKING

For those interested in making candles at home, craft and hobby shops often stock candle molds, paraffin, beeswax, stearic acid (used in paraffin candles to raise the melting point and subdue smoking), colors, and scents. My friend Gordon Pine likes beeswax candles, but since he burns a dozen a week it gets expensive. So he and his wife economize with a beeswax scented paraffin candle. Here's his recipe:

- 1½ lbs. paraffin
- 3 small chunks of beeswax (about half-thumb size; this is a minimum. One can certainly add more. I add beeswax because even a little gives it a better consistency, a better look, and a better feel.)
- 3 tablespoons stearic acid
  Liquid scent and color (cake type) as desired
  Prewaxed wick

Melt paraffin and beeswax in a double boiler, add stearic acid, and color/scent as desire. We use three to six two-stick tin molds for 10-inch candles. Pour in mold, let cool, and remove. Makes about 36 candles.

Classic all-purpose holder, this candle pan is sometimes called a "saveall." It's handy for burning stubs that can be collected in the pan. The reflective brass carrying handle is perforated so it can double as a wall bracket.

For outdoor burning, or anywhere in a breeze, a glass hurricane chimney is necessary to protect the flame. A blowing flame gutters, drips, and burns unevenly.

Just about any candle may drip now and then. To catch running wax use a drip dish, or *bobeche*. You can make one by inserting a disk of aluminum foil in the candle holder so a collar of about a half inch extends beyond the base. Glass *bobeches* are available too. Candles should not be allowed to burn down to the stub in glass or ceramic candlesticks, as the candlestick is likely to break.

Around the wick at the top of a burning candle is a small pool of hot liquid wax. When you blow out the candle, be careful not to spatter hot wax around. Better yet, extinguish the flame by capping it momentarily with a snuffer. If your local merchant doesn't stock any, try a spoon.

Finally, if you don't like the lingering smoke of a freshly extinguished wick, wet your fingers and give the wick a quick pinch. I enjoy a brief whiff of beeswax candle smoke, but not that of paraffin.

The Kerosene Wick Lamp

Not long after moving to the hills I was invited to a neighbor's for a night of music. I'd never been to Martha's and I looked forward to some music and dancing and to seeing her place. It was an old farmhouse, weathered and gray. Dusk was settling when I arrived. I'll never forget the vision as Martha swung open the door. Behind her, covering a round oak table, flamed a constellation of kerosene lamps. There were glass lamps, each one polished diamond bright and filled to the brim with fuel. The room was lit by an orange rainbow. Pumpkin-red flames blossomed over glowing lamp reservoirs, and everyone's face seemed bathed by a Halloween bonfire. Martha had just polished and filled the lamps, and I helped disperse them throughout the house. It was a weekly ritual, she explained, one of the ceremonies of living without electricity.

A while later I cut the electric cord myself. Some of my friends contemplating a backwoods move balked at living non-electrically. I didn't hesitate. That night at Martha's convinced me.

It's been a decade since then, the early years illumined solely by kerosene light. Even after I replaced most of the lamps with brighter liquid propane (LP) gas mantle lights I saved several wick burners, and not for nostalgia's sake. Unlike LPs, wick lamps are portable, their light level can be adjusted, and I don't have to worry about broken mantles. I've used kerosene lamps to thaw out frozen water pipes and keep the root cellar from freezing on thirty-below nights. And if the LP runs out, kerosene lamps are a must. Best of all, they're inexpensive; most hardware stores carry simple kerosene lamps costing anywhere from $7 to $15 (of course, fancier models and antiques can cost a lot more).

The kerosene wick lamp is the granddaddy of firelight. It hasn't changed significantly since the aerodynamically tapered glass chimney and vented burner were devised more than 150 years ago; before that, the best lamps were little changed from the fat-burning stone cressets of 20,000 B.C. What makes the wick lamp—from the Greek *lampein*—such a masterpiece of timeless design? Simplicity. The ubiquitous, inexpensive wick lamps at the local hardware store and the imposing European antiques all have four basic elements in common: fuel reservoir, burner, cloth wick, and glass chimney.

A wick lamp works quite simply. Fuel in the reservoir, or font as it's sometimes called, is soaked up by a wick that is ignited at the burner. The resulting flame is both protected by the chimney and brightened by it, as the chimney creates a draft. There are some minor subtleties to the process, which we'll get into as we go over a lamp, part by part, and discuss what fuels it. The flames from lamp combustion deliver light equivalent to that of a twenty-five to forty-watt bulb, and their illumination can be amplified by using a reflector, or a double-wick burner. The light may not be sufficient for detail work, although in a pinch, it's enough for reading.

Fuel

Kerosene is the standard fuel for wick lamps. The term kerosene is used loosely to describe a thin flammable oil with a rather high ignition or flash point, roughly 160°F. That high ignition temperature makes kerosene safer for household use than more volatile alcohol or gasoline—in fact, a lighted match can be dunked in a pot of kerosene without igniting it.

Kerosene can be produced from coal oil, oil shale, and wood fiber, but it is refined from petroleum most commonly. It can also be made in different grades, so not all kerosene is created equal.

What you want is a special low-sulfur kerosene, usually referred to as Number 1-K. It is also sometimes called Type A, paraffin

Kerosene lamps are distinguished by their wicks and fuel reservoirs. Left to right: a single wick with transparent glass reservoir, a duplex wick with ceramic reservoir, and a round wick with brass reservoir.

oil, or water-clear or lighting-grade kerosene. This is the champagne of kerosene, without the bubbles. Number 2-K kerosene, which is a lower grade and much less expensive, can also appear clear, but it can contain almost ten times as much sulfur as 1-K—0.3 percent by weight (maximum) versus 0.04 percent for the purer stuff. Also, be careful not to purchase other fuels that may be passed off as kerosene—fuels such as Number 1 fuel oil, Number 1 diesel fuel, and jet engine fuel. These fuels may have a telltale tawny color and sour smell.

The trouble with using fuels other than 1-K is that they contain too much sulfur and tar. Gradually, the tar gums up the wick, reducing capillary action. The more low-grade fuels you burn, the dimmer the light and the smokier the flame. If you're compelled to burn 2-K, for instance, it's necessary to clean the wick frequently—or replace it—and also clean the reservoir. That still won't increase the illumination, however, or reduce the unpleasant pungent odor when it burns. A wick lamp is not fuel efficient; it burns only about 70 percent of the fuel, and if the unburned kerosene vapor is loaded with sulfur, you smell it. Switching to 1-K won't boost efficiency much, but it sure cleans up emissions. In upgrading lamps to 1-K, be sure to install new wicks and scour the bowls; otherwise, pouring in 1-K would be as fruitless as trying to keep cider in a vinegar barrel.

Lighting-grade fuel is sold under a variety of brand names. I happen to use TruLite, which runs about $4.60 a gallon. Quart bottles are more expensive, per unit. Be careful of scented or colored lamp oils, which are often 2-K in disguise. If you're lucky, you can find 1-K kerosene to buy in bulk at local oil suppliers or at hardware stores or other retail outlets that sell portable kerosene heaters, most of which work best with 1-K fuel. From these sources, 1-K sells for between $1.20 and $1.50 a gallon, in your own containers. (Kerosene should be stored in an airtight metal or petroleum-safe plastic container in a cool, well-ventilated place. It will begin to congeal at $-30\,°F$. If it congeals, never thin it out with another flammable substance. Bring it inside to thaw.) Because one lamp will burn roughly forty hours on a quart of kerosene, the cost with the higher-priced fuels comes to less than $ .03 an hour per light. And if you can buy your kerosene by the five-gallon can, you'll pay less than a penny per hour. (The rate will vary with different wicks.)

Reservoirs

Of course, there's more to lamplighting than simply burning 1-K kerosene. In choosing the lamp, for example, you can select from several styles of fuel reservoir. The reservoir may be glass, metal, or ceramic. Glass has the most advantages—and liabilities. A glass bowl offers a constant reading on the amount of fuel and the length of the wick. That's helpful, because it's important to maintain at least a half-inch of fuel in order to sustain bright illumination. It's important also to keep a generous length of wick in the bowl for good capillary action. If you're burning several lamps with any regularity, you'll appreciate not having to open up the bowls to check the fuel or the wick. A glass bowl also lets some light through to the tabletop, while a metal or ceramic bowl casts a shadow.

On the other hand, if the glass reservoir breaks you're in trouble: a big mess, or a fire. Still, I prefer glass bowl lamps, perhaps because I've been lucky enough to avoid breakage. Others haven't been as fortunate. According to a report by the Consumer Product Safety Commission on accidental lamp fires, overturned lamps are the major source of injuries.

You can eliminate the risks of glass with an unbreakable metal reservoir. Solid brass, aluminum, or steel coated with enamel, nickel, or brass are used for lamp bowls. One of my first lamps was a nineteenth century round-wick Vestal with a silver plated, one-quart reservoir. It didn't take me long to catch on to the advantage of the ample fuel capacity. A big reservoir reduces the need to refuel, and that's particularly useful when you can't see how much fuel you're burning. I also like the Vestal's separate filling hole, a handy feature on any reservoir. A separate hole makes it easy to use a dipstick for gauging the fuel supply, and simplifies refueling too. Otherwise, reservoirs must be filled by removing the burner and chimney assembly, including the wet wick. However you fill the reservoir, be careful not to overfill it; kerosene expands somewhat as it warms up, and you don't want the lamp to overflow.

I know potters who make exquisite lamp bowls, and indeed many ceramic lamps are available. But I can't see the point of a fragile reservoir that doesn't offer the viewing advantages of glass. If you do happen to chance upon an irresistible ceramic lamp, be sure to check the finish. One Christmas I received a clay lamp that burned brightly that night, but next morning was seeping kerosene onto the table. The potters had neglected the interior glaze.

Burners

Plugged into the top of the fuel pot is a dome-shaped device that brings all the elements of the kerosene lamp into focus: the burner. The burner houses the wick and its height-regulating mechanism, supports the chimney, and holds the flame. Moreover, when the lamp is lighted, the burner draws preheated air through a circular opening just below the flame, thus enhancing combustion. Together with the chimney, the burner transforms an inherently dim, smoky flame into a bright, clear light. And on many lamps, as I said, the burner also serves as the cap to the fuel reservoir.

Burner sizes differ according to the size of the reservoir collar, the chimney base, and the wick width. A Number 1 burner fits a 7/8-inch collar, holds a 2½-inch chimney, and takes a flat 5/8-inch wick. A Number 2 burner fits a 1¼-inch collar, holds a 3-inch chimney and a 1-inch wick. Smaller "Nutmeg" burners fit Number 00 collars and take a 3/8-inch wick and a smaller $1^1/_8$-inch chimney. To protect against corrosion, burners are made of solid brass or brass-plated steel, and are simple to maintain. The kerosene keeps the wick regulating mechanism oiled, and the dome is often hinged for easy access to the wick when it's time for trimming or replacement. With the dome tipped up, you'll see that the circular opening that provides the draft is screened. That keeps wick ashes from falling through the burner. The screen should be cleaned periodically to allow unrestricted air flow. Avoid cheap burners with no protective screening. And make sure not to bend or crimp the wick shaft; the shaft must allow the wick to turn up smoothly to produce an even flame. (One type of lamp, made in England and called the Duplex, is fitted with a two-wick burner. The Duplex is highest in candlepower of the flatwick lamps I've seen.)

Wicks

The wick and its turnup wheel are the lamp's only moving parts. The wick must fit snugly in the wick mounting, yet be free enough to move and draw a copious flow of kerosene to the flame by capillary action. Over the years, a flat ribbon of woven cotton has proven the easiest to maintain. Tubular wicks, like those used in European imports and Aladdin mantleless lamps, are a bit tricky to trim evenly, and it's no cinch to devise a regulating mechanism that lifts the wick symmetrically. The Aladdin round wick is not foolproof, but the European Kosmos is reliable and very bright.

Installing a lamp wick is easier than threading a needle, but not much. It may help to cut a fresh edge on the wick before slipping it into the mounting shaft in the burner. As the wick is fed in, the regulating wheel should be turned until the wick catches and emerges. In particularly tight shafts, it may be necessary to wet the wick in kerosene before fitting.

The lamp burns best when there is roughly 1/8 inch of fiber topping the shaft—and 3 or 4 inches of wick at the bottom of the fuel pot. The lamp won't burn at its peak candlepower if the wick has only a shallow rooting in the fuel. Besides, you'll have to fill the reservoir constantly just to keep the wick saturated.

Unlike a candle wick, which is woven so that the fiber is consumed in the flame, the lamp wick is not self-trimming. After twenty or thirty hours of normal use, the wick grows charred with the residues of unburned fuel, and the flame dims and smokes. It takes just a moment to extinguish the lamp and cut a fresh edge. To produce an even flame, chop the wick straight across with a pair of scissors (it sometimes helps to cut off the corners first). After trimming, clear any charred remains from

Duplex (double-wick) burners offer single- or double-wick burning, thus the broadest illumination spectrum in kerosene wick light. Many double wicks have a mechanical snuffer, especially handy on overhead lamps.

the burner screen. I know some people who prefer to clean the wick by simply rubbing off the charred edge with their fingers, or a match, which saves time. To minimize charring, watch that the flame is not set too high. Turned up beyond an efficient level, the lamp's alarm goes off: the flame turns red and begins to smoke.

When you see that it's time to replace the entire wick, remove the burner and use the regulating wheel to eject the old wick.

Chimneys

Topping off the wick lamp is the glass chimney, and like other elements of the lamp it performs several important tasks simultaneously. The chimney creates the draft that increases combustion and brightens the light, and it's also a wraparound windshield to protect the flame.

Straight-sided glass chimneys were first used on wick lamps in the eighteenth century. Before the close of that century, Aime Argand, a Swiss inventor, with help from a French assistant, discovered that constricting the chimney just above the flame aided combustion and amplified the light. Hence the present-day chimney's graceful taper above the bulbous base. Unlike the hurricane chimney sometimes used to shield candles, the wick lamp's chimney is a necessity. Without it, the lamp would be smoky and dimmer than a candle.

Standard chimneys are made with clear, frosted, or patterned glass and come in several sizes: Number 00 with a $1^{1}/_{8}$-inch base or "fitter;" Number 0, with a $1^{5}/_{8}$-inch fitter; Number 1, with a $2^{1}/_{2}$-inch fitter; Number 2, with a 3-inch fitter; and Number 3, with a $3^{1}/_{2}$-inch fitter. The largest chimneys, called Store Lamps, have 4-inch fitters.

Chimney heights vary, with $7^{1}/_{2}$ inches to 12 inches being most common. Above an altitude of four thousand feet, where extra oxygen is needed for combustion, many lamp keepers prefer even taller chimneys.

Chimney glass is tempered for heat resistance, but there's no getting around its fragility. Held in place at the base by an expandable metal ring or several clamps, the chimney is easy to slip off for cleaning, wick trimming, or burner removal; and it stands secure when the lamp is stationary. Nevertheless, the clamp cannot always be counted on when the lamp is being moved. If need be, steady the chimney by holding it at the base, where it's cool enough to handle.

A lamp that is properly fueled and maintained won't often need chimney cleaning. When it does, a little soap and water is sufficient. To prevent breakage caused by thermal shock when cleaning, be sure to let the chimney cool before dunking it in water. Apart from that, it's possible to dry clean the glass by twisting a wad of newspaper inside the chimney.

When it comes time to blow out the lamp, aim for the chimney top. Cup a hand on the side away from you for best snuffing. To curtail smoking, turn the lamp down low before extinguishing. (The same is true for ignition: let the lamp warm up for a moment before raising the flame.)

Safety Tips and Accessories

Selecting a safe site is perhaps the lamp keeper's greatest responsibility. Sometimes it's not easy to find a spot for the lamp that is both effective for shedding light and out of harm's way. On a kitchen counter, for instance, it may be difficult to light up the sink and also keep the hot chimney protected from splashing water, which could crack the glass. A workbench, where chips are flying, poses a similar threat. Often a lamp bracket solves the problem. Brackets of decorative cast iron can be mounted on wall or post, yielding an extra margin of safety. Lamps designed for hanging are also available. These lamps often include a chimney cap to deflect heat, making them well suited to hang from the ceiling. Remember to follow the lamp maker's specifications for safe clearances to the nearest combustible surface.

Anyone on the lookout for shipboard kerosene lighting will be pleased to hear that gimbal-mounted wick lamps are still made. These lamps remain vertical even when the hull doesn't.

The ability to adjust illumination is one of the highlights of the wick lamp, and the lamp keeper can further extend the light range by using reflectors and shades. Tin-plated steel reflectors or mercury dish mirrors amplify illumination without burning a drop of extra kerosene. Those who prefer their flames tinted can slip colorful glass shades over the chimney to be supported by a brace on the burner.

Aladdin Light

Bright as a sixty-watt bulb, with adjustable illumination, portable, available in more than a dozen models with better than a dozen shades, and adaptable to electricity: this is the Aladdin, the most versatile kerosene lamp in the world. With a few refinements, it's the same lamp that revolutionized oil lighting three-quarters of a century ago. Today, fifteen million Aladdins later, it is still the Aladdin that many lamp keepers turn to for non-electric lighting. There's no firelight available that burns as brightly without the aid of a pump or pressurized tank. And there's no other lamp as sneaky. If you don't keep a close watch on the Aladdin during warm-up, it can erupt like a volcano, spewing smoke through the house, blackening the mantle with carbon, and perhaps touching off a fire. This is one dimension of the lamp that advertisements don't mention. The reason so many people live with this peril is clearly because Aladdins give a high quality of light when everything is working right.

Lighting with Style

"Aladdin light" is a phrase that describes a special quality of illumination as well as the lamp itself. This is light with a character all its own. The turnup wheel can be spun from dim to dazzling, in contrast to most mantle lamps of fixed candlepower. Moreover, no matter what the setting, the incandescent mantle diffuses the light with an effect similar to that of a frosted light bulb. It is soft, with an ultramarine luster, a light I've heard described as "deep." And although the light is triggered by a cotton wick burning beneath the mantle, the Aladdin has none of the flicker characteristic of wick lamps. The mantle "freezes" the flame. Flow lines of illumination beam out in a 360-degree flood, although the Aladdin does cast a fuel pot shadow.

The Aladdin is a hybrid. It combines two unique elements of firelight design: the tubular wick and the Welsbach mantle. The tubular cotton wick was devised by Aime Argand, working in France in 1784. It was a terrific advance over the flat ribbon wick, although it proved difficult to perfect. Argand's wick consisted of a ring of cotton wrapped around a central draft tube. The snags involved designing a turnup mechanism that would consistently deliver a symmetrical flame, and solving problems of trimming—troubles that can still pester Aladdins today. Nonetheless, when it worked, Argand's wick delivered ten to twelve times the customary light from a fuel lamp.

One hundred years after the appearance of the Argand lamp, Austrian chemist Carl Auer von Welsbach invented the incandescent lamp mantle. This is a small gauze pouch impregnated with oxides of the rare-earth metals thorium and cerium. (This is further described in the "Incandescent Mantle" section.) Heated in a flame, the mantle yields a brilliant light because of the incandescence of the oxides. Little more than a brittle ash, the mantle will not incinerate below the melting point of the materials: 1750°C. In many lamps, the Welsbach mantle is ignited directly by a gas flame (it was discovered over a Bunsen burner and today shows up also on LP gaslights and Coleman-type gas pressure lamps). However, in the Aladdin, the mantle is ignited by the flame of an Argand-like tubular cotton wick. Hence the unique character of the Aladdin: it's a kerosene lamp that burns as bright as pressurized gas—or an electric bulb.

At the turn of the twentieth century, the tubular wick and the Welsbach mantle were first combined in a German lamp called the Practicus. Imported to the United States, the Practicus caught the eye of a young Nebraskan, Victor Samuel Johnson, who saw its potential for rural sales. In 1907, Johnson formed a company in Kansas City, Missouri, to import and sell the Practicus. Soon there was sufficient business to prompt Johnson to move his outfit to Chicago, where he began work on improving the lamp. To eliminate smoking and unreliability, he acquired a patent for a new lamp burner from an inventor named Charles Wirth, and, in 1909, Johnson introduced the first Aladdin lamp. He named it after the mythical lad in the *Arabian Nights* who won a fortune from a genie in an oil lamp. By 1915, after more improvements in the burner, mantle, and flame spreader, the Aladdin was winning gold medals at international expositions, and Johnson was on the way to his own fortune.

Today, Victor Samuel Johnson, Jr. presides over the company that his father founded, selling a lamp that has changed more in style than in substance. The key elements of the Aladdin—the tubular wick and the incandescent mantle—differ little from their turn-of-the-century predecessors. However, the lamp bowl has taken multiple shapes, and many of the early glass models, especially the crystal table lamps made during the thirties, are now valuable collector's items.

The lamp consists of four basic parts: the reservoir, the burner (which incorporates a tubular cotton wick), the mantle, and the chimney. Aladdin lamps are all quite similar from the burner up (with the exception of various shades). The most significant options involve the reservoir.

The Reservoir

The Aladdin reservoir is produced in a variety of shapes and materials. In fact, the dozen or so different reservoirs have only one thing in common: fuel capacity, which is one quart, or roughly twelve light hours. Reservoirs are made of aluminum, brass, clear glass, and porcelain. There are low reservoirs for the desk, table model reservoirs on five-inch pedestals, and bowls that can be shifted from table to wall bracket to suspension hanger.

Most people prefer a metal reservoir because it's unbreakable. It's particularly wise to choose metal over glass in the case of Aladdins, because they tend to be more top-heavy than traditional kerosene wick lamps; the pedestal models crowned with a glass shade are especially so. A metal reservoir is cheap insurance against accidental breakage and a flammable oil spill. Reservoirs of brass or less expensive aluminum are available in the traditional

Aladdin styles: the pedestal table model and the desktop. The low boy desk lamp is designed to adapt to a wall bracket or suspension hanger as well, making it the most flexible model.

Aladdin reservoirs are also available in clear and tinted transparent glass and opaque porcelain. Transparent glass bowls are made in both desktop and pedestal styles. The advantage of glass is that the lamp keeper can check the fuel supply at a glance. Because it's especially important to keep the Aladdin from running dry to prevent damage to the wick, a lamp reservoir that doubles as a fuel gauge is a significant convenience. The tradeoff, of course, is that the glass is fragile.

Porcelain reservoirs are attractive, although they seem the worst of both worlds: a fragile bowl that hides the fuel. Even more vexing, most porcelain bowls lack a filling cap. That means the lamp keeper must unscrew the burner from the bowl to fill up, which exposes the wet wick, and special care must be taken to avoid spills. (A filler cap might not be as important on a wick lamp that needs a refueling every forty-eight hours or so, but the Aladdin burns kerosene four times as fast.)

Because the lamp reservoir accounts for so much of the Aladdin's function, it's easy to customize a lamp by simply screwing on a different reservoir. For instance, you might find a secondhand Aladdin with a pedestal bowl you don't like, or one that lacks a filler cap. For a modest sum you can order a bowl better suited to your lighting requirements.

The Burner

Like the carburetor on a combustion engine or the draft on a wood stove, the Aladdin burner mixes air with fuel to make fire. In the engine, fire is converted to horsepower. In the stove, fire is concentrated for thermal power. In the Aladdin, fire is converted to candlepower. The burner makes it possible for the Aladdin to generate light from a highly efficient mixture of 6 percent kerosene and 94 percent oxygen. Thus, even though you may burn low-grade kerosene, the Aladdin is nearly odorless. (A wick lamp throws off 30 percent of its fuel as unburned vapor; the Aladdin wastes hardly a drop.)

Using the flame of a tubular wick to heat its incandescent mantle, the Aladdin bridges the gap between the classic kerosene wick lamp and the modern mantle light.

Threaded into the lamp reservoir, the burner sits atop the fuel supply and supports the mantle and chimney gallery. The burner consists of a perforated metal cyclinder, a tubular wick and its regulating mechanism, and a flame spreader. Burner design is uniform on all late model Aladdins, with the option to choose between solid brass and a brass alloy coated with chromium.

The wick draws up fuel by capillary action. At the top of the burner, the wick is exposed about an eighth of an inch. At the top of the burner, the wick burns with the help of a double draft feeding up through the perforated burner wall. Air rushes to the tubular wick from inside through the flame spreader and from outside through a perforated ring that encircles the wick. As the wick flame warms up, the mantle begins to glow. The hotter the fire, the brighter the light. Unlike a traditional wick lamp, the Aladdin wick burns to make heat, not light.

Today's Aladdin burner is a sophisticated descendant of its original Practicus counterpart. The Practicus burner consisted of a cylindrical wick with a draft tube through the center of the fuel reservoir. The reservoir fit above a pedestal base. At the base of the pedestal was a ring of small vents to draw up air. Adapted to the Aladdin, this center-draft burner remained in place until the 1930s. But there was a problem. The light took a quarter-hour or more to stabilize, because air passing through the full length of the reservoir pedestal warmed up so gradually. Hence the lamp keeper had to monitor the Aladdin carefully after ignition. Otherwise, as the temperature rose, the lamp would overheat and flare up.

In 1934, Aladdin introduced the side-draft burner. This new design eliminated the draft tube through the lamp bowl. Instead, the burner drew air through a perforated cylindrical wall above the fuel reservoir. With less mass to steal heat, the Aladdin reached peak illumination in a couple of minutes. The new lamp became known as the "instant light." The side-draft lamp was safer, and because the new burner drew more air to enrich combustion, it brightened the light. The center-draft Aladdin had generated eighty-five candlepower; the instant light boosted candlepower to 125. The new burner also gave the Aladdin designers greater flexibility. The center-draft lamp had been limited to pedestal models for table use only. But with the side-draft burner, Aladdin reservoirs could be set flat on a table, and moved from desk to wall bracket to pedestal, without disturbing the draft.

At the center of the burner is the tubular wick. The wick consists of a tightly woven cotton tube dangling a pair of cotton tails of a looser weave. The wick fits into the burner so that the tails are immersed in the kerosene while the top is encased in a tubular metal sheathing. The wick is adjusted by means of a rack and pinion mechanism. The rack clips to the base of the tubular wick, which is raised and lowered by the geared wick button. Since 1922, the wicks have been manufactured with an inner and outer reinforcement to prevent deformation and to hold a symmetrical burning edge. The wick's edge is beveled to facilitate installation and cleaning; the wick itself will burn for about a thousand hours.

The flame spreader is the thimble-like cap that fits into the top of the draft tube. Its concave, perforated top deflects air evenly to the surrounding flame. The flame spreader also acts as a protective top on the draft tube, preventing ashes from falling into the burner. During wick trimming, the flame spreader is removed.

The burner is topped by the gallery, a circular twist-on metal ring. Both the mantle and the chimney fit into the gallery, which enables the lamp keeper to remove them for lighting and wick trimming.

The Mantle

The mantle is approximately three inches tall, conical, and resembles a spider's web. It is the heart of the lamp, the source of light. The mantle's rare-earth filaments make it possible for an unpressurized kerosene lamp to produce four to seven times the illumination of a customary kerosene lamp.

The mantle is extremely delicate and must be handled with care. In fact, it should not be handled at all. The mantle is framed in a metal structure so you can install it without touching the brittle filaments.

The mantle is the product of an interesting manufacturing process. It begins as a tubular open-weave rayon knitting. This is saturated in a solution of thorium and cerium nitrate salts. After drying, the mantle is burned to remove the rayon, and intense heat is applied to change the nitrate to oxide. What remains is a delicate metallic web of rare-earth oxides: 99 percent thorium and 1 percent cerium. Finally, the mantle is coated with lacquer for protection during shipping and installation. After fitting the mantle on the gallery, the lamp keeper burns off this lacquer. Henceforth, whenever the mantle is heated by the Aladdin wick, the filaments glow.

The Aladdin mantle is unique. It is roughly double the size of a Coleman or liquid propane lamp mantle. And instead of hanging down under a pressurized flow of flammable vapor, the Aladdin mantle stands upright over the circular wick. Moreover, as part of the gallery, it is the only lamp mantle I know that must be removed temporarily during ignition. Consequently, the Aladdin mantle is subject to more stress than any other. That's why the Aladdin mantle is reinforced by a metal brace; that's why the mantle is more expensive than others; and that's why, unless the lamp keeper is careful, the mantle frequently will break.

The Chimney

The Aladdin chimney is easy to recognize. At 12½ inches, it is taller than most lamp chimneys. With only a slight outward curve to accommodate the mantle, it looks more like a glass smokestack than the traditional bell-shaped tapering lamp chimney. This sleek design represents a pinnacle in firelight aerodynamics: the high-velocity draft chimney makes it possible for the Aladdin to burn 94 percent oxygen and 6 percent kerosene, producing 125 candlepower. No other unpressurized liquid fuel lamp delivers such brilliant light.

The chimney fits tightly into the gallery to insure a consistent draft. When it's time to ignite the lamp, chimney and mantle are removed together as part of the gallery and replaced after lighting. The glass is heat resistant, of course, but it doesn't have much resistance to thermal shock. It's a good idea to situate the lamp so that no water splashes on the hot chimney.

It's also important to give the lamp adequate overhead clearance for proper combustion and to insure against overheating and fire. An open chimney should be sited at least thirty-six inches below any material, combustible or otherwise. Aladdin makes a couple of smoke bells that can be used over the chimney to deflect heat. One smoke bell fits the hanging lamp, about four inches above the chimney. Another smoke bell is designed for wall-mounted lamps. Nevertheless, despite the protection of the smoke bell, be sure that the chimney top is no closer than thirty inches to an overhead surface.

Another optional chimney fitting is the insect screen. This is a perforated chimney cap to protect the mantle. There's little chance that a bug is going to dive down the chimney of a burning lamp. But when the lamp is out, bugs may wander down and break the brittle mantle.

Aladdin also manufactures chimneys for high-altitude lamp keepers. A healthy flame needs a good diet of oxygen, and at high altitudes a taller chimney compensates for the thinner air by increasing the draft. This is especially important in the Aladdin because of the high-oxygen/low-fuel carburetion. Aladdin's high-altitude chimney is suggested for use at elevations over four thousand feet. It's 14 inches tall, rather than the standard $12\frac{1}{2}$. Yet for some lamp keepers at especially lofty altitudes even the high-altitude chimney isn't enough. The lamp is slow to warm up, never yields full illumination, and the mantle carbonizes. Nevertheless, Ruth Young of Aladdin customer service tells me that it's possible to burn their lamps in the mountains without much trouble.

"Oh yes," she assured me, "We have people using Aladdins up on Pike's Peak all the time. Aladdin used to offer a metal chimney extension, about five or six inches, but it's been discontinued. Now most people make one themselves, or have a tin shop do it." Mrs. Young added that orange juice cans, with both ends removed, used to work fine. But nowadays they're made of cardboard, so that's out. I wonder still if there isn't a soup can or something that might do. In a pinch, a few layers of

An Aladdin lamp can be at home in many locations—on a suspension hanger, a wall bracket (as shown)), and a table. There's also a pivoting wall bracket for use in a boat or camper.

aluminum foil could be fashioned into an extension. Whatever the non-combustible material, the extension should be designed to slip over the chimney and rest on the glass where it expands below the rim.

Assembly and Lighting

Setting up a new Aladdin is a matter of assembling the four basic lamp elements: reservoir, burner, chimney, and mantle. To begin, the burner is screwed into the reservoir. (If the lamp includes a tripod shade holder, first it must be attached to the lamp between the burner and bowl.)

The mantle is then installed in the burner gallery. Grasp the mantle by the wire frame; do not touch the mantle fabric. The mantle is locked in place by turning it clockwise. The chimney also is fitted to the gallery with a clockwise twist. Be careful not to overtighten the chimney. The base should fit snugly, but with enough leeway to expand and contract with the heat.

Now the lamp should be filled with a good grade of kerosene—preferably Number 1-K or Aladdin paraffin oil. Never use gasoline, alcohol, or other dangerous fuels. Neither is the use of scented lamp oil recommended, because additives in the fuel will foul operation of the lamp. Allow the wick to soak up the kerosene for at least one hour before ignition; the wick must be completely saturated to burn correctly.

Next, burn off the mantle's protective lacquer. Remove the chimney and hold a lit match near the bottom of the mantle, but do not touch the fabric. The protective coating will burn off in a flash. To keep from smoking up the house, and your lungs, do this outside whenever possible, making sure the mantle is protected from the wind. Replace the chimney.

When you want to light the Aladdin, remove the gallery, mantle, and chimney as one unit. Turn up the wick about 1/8 inch above the outer wick tube, and ignite with a match. Allow the flame to cover the entire top edge of the wick, and then carefully replace the gallery assembly and lock it into place. Slowly turn up the wick until part of the mantle glows white. Let the lamp warm up for several minutes. Never turn up the lamp to

The "gallery" holds the mantle (shown here installed) and the chimney, which can be removed for cleaning. The gallery, with mantle and chimney in place, must be removed each time the lamp is lit. Caution: When you are relighting after short intervals, the gallery may be hot.

*If a flareup occurs, do **not** sprinkle salt down the chimney, as many oldtimers recommend. The salt will corrode the burner. Instead, turn the lamp down low and allow the flame to erase the carbon slowly.*

brightest light immediately because this will cause the lamp to flare up and smoke. After the burner has warmed up fully, the heat generated will cause the mantle to increase the intensity of its glow. Correct lighting is attained with an illuminated mantle that displays no points of orange flame breaking through the mantle. If flames appear, the wick should be turned down.

To extinguish the lamp, turn the wick down just below the wick tube until the flame disappears. Blow softly into the burner or across the top of the chimney. Then raise the wick to see that the lamp is out. If you want to light the Aladdin again soon, beware. The gallery will be hot.

The Aladdin requires fresh air, and so does the lamp keeper. Best results will be obtained in well-ventilated areas. Never use a lamp in a sealed area where oxygen is short and carbon monoxide fumes can gather.

It's important to maintain safe distances between the lamp and any combustible surfaces or materials. A clearance of at least thirty to thirty-six inches above the top of the chimney is recommended by the manufacturer. Always allow adequate space around the lamp to prevent excessive heat buildup.

If you are using a shade, be careful during warm up; the mantle will be invisible. In addition, handling a shaded lamp may be awkward during ignition. Remove the gallery with care.

Maintenance

The Aladdin must not be allowed to burn dry. This would damage or destroy the wick. The flame spreader must be kept clean. If it becomes dented or damaged, replace it. If the mantle becomes blackened from soot or carbon, it should be cleaned by turning the flame down low and letting the crust burn off slowly.

The wick should be cleaned only if it has a formation of carbon crust on the edge. To clean, remove the gallery assembly and turn down the wick until the top edge is even with the wick tube. Remove the flame spreader, and insert the disk-shaped wick cleaner that comes with the lamp into the wick tube. Turn up the wick until it presses gently against the cleaner. Now slowly turn the cleaner clockwise until the wick is smooth. Do not ruffle or gouge the edge, because a ragged wick will produce an uneven flame.

Lantern Light

Firelight is a natural for outdoor illumination, but, as you may have discovered, not just any lamp will do. Take a table lamp outside and you're likely to find it flicker, smoke, and die in the breeze. These lamps are designed to burn in still air where the chimney draft is steady and undisturbed. And if you've ever tried to navigate a dark trail with a table lamp in hand, you'll know how difficult it is to see beyond the glare of the flame. Moreover, an uncapped chimney is open to precipitation, and foul weather will extinguish the lamp or perhaps crack the hot glass. Glass bowl lamps are fragile besides, and risky for outdoor use.

Outdoors and Portable

Wherever there's a need for outdoor portable light, you're likely to see a fuel-burning lantern—either a high-powered mantle lantern or a simpler kerosene wick burner. Battery-powered flashlights and lamps are fine for short-term use, but nothing beats a lantern for bright, inexpensive, stormproof illumination, whether for camping or general purpose use around the house or farm. A gas or kerosene pressure lantern will generate more than 200 candlepower—at least the equivalent of a 100 watt light bulb—and burn for eight hours per filling. A kerosene wick lantern will generate up to twelve candlepower and burn for approximately forty hours each filling.

A lantern is a miniature lighthouse. In a kerosene wick lantern, the weatherproof draft system draws oxygen to the wick flame, and the exhaust rises through a perforated, lidded chimney top.

The Coleman lantern is a little lighthouse. Within a weathertight enclosure, its twin mantles generate up to 220 candlepower. And better than a lighthouse, it's portable.

A gas or kerosene pressure lantern with mantle features a similar protective chimney, although the brighter mantle light is fed by a vaporous combination of compressed air and fuel. (LP gas cartridge lanterns are available also, but I find them expensive to operate and without much merit; they operate on the same principle as LP gas lights.) Both wick and mantle types incorporate glass chimneys specially tempered for resistance to impact and temperature shock. The lantern body is unbreakable steel. A hinged "bail" similar to a bucket handle enables the lamp keeper to carry the lantern below eye level, so that light falls on the terrain without any blinding glare.

The Pressure Lantern

The name Coleman is synonymous with pressure lanterns. Since 1905, the Coleman Company has been producing lanterns in its Wichita headquarters; currently they sell a million or so annually. The company makes several lantern models, including bottled propane lights, but they are known primarily for the quick or instant light pressure lanterns, which feature a hand pump on the fuel tank and burn kerosene or Coleman gas, an unleaded gasoline sold in the familiar red cans found in most camping and general supply stores. Like the Aladdin table lamp, the Coleman pressure lantern is based on a patented design, and it is unique in its field.

The first Coleman lanterns appeared at the turn of the century to compete with several emerging illumination systems. At the time, kerosene, natural gas, and manufactured coal gas fueled a new generation of indoor lamps featuring the Welsbach mantle, and Edison's electric light was catching on. But the mantle lamps were limited to a maximum output equivalent to approximately seventy-five watts per unit, and early electric light bulbs barely exceeded twenty-five watts. An empty spot at the high end of the spectrum beckoned: a powerful light for street, store, church, meeting hall, and lighthouse, as well as for farmers and others who worked outdoors.

The Coleman lantern filled that spot. The first Colemans burned a pressurized gasoline vapor produced by hand pumping an air compressor and then preheating a vaporizing generator with an alcohol torch. Once ignited, the gasoline vapor heated an incandescent Welsbach mantle that provided the light: the heat also sustained the vaporizing process.

What caused the Coleman to be so much brighter than other lights? Primarily it was the built-in air pump, which pressurized the pedestal fuel tank to roughly forty pounds per square inch, creating a fuel-injected light. Combined with the vaporizing effect of the pressurizer and the highly volatile gasoline, the Coleman was quite simply the hottest lamp around—and the higher the heat, the brighter a mantle would shine. Early Colemans delivered approximately 750 candlepower of bright white light from a durable ten-pound unit that could stand on its pedestal tank or hang suspended. In 1905, Coleman lamps illuminated the first night football game, in Wichita. The light was so effective that some people complained it was too bright for household use.

By 1914, Coleman was producing the world's first stormproof pressure lanterns, lit by double mantles. Toward the end of World War I the company introduced its first "Quick Light," incorporating a new generator that could be match lit, quite an improvement over alcohol torch ignition. The lantern grew especially popular with merchants and farmers, and during World War II "Coleman" became part of the language for a whole generation of Americans—the GIs. Coleman devised a pocket stove that became an essential ingredient in many GI packs, and the company received a special government award for its contribution to the war effort.

Indeed, there's something about a Coleman that reminds me of battle. It was at a barn dance on the fourth of July, years ago, that I lit one for the first time; upon ignition, the lantern flared up like a Roman candle. A plume of fire shot up and singed my eyebrows before I ducked and shut the valve. Then a funny thing happened. The lantern began to glow. Hesitantly I reopened the valve. Light! That night a single lantern illuminated the cavernous barn, and after midnight we carried it outside, despite the rain, to help get our fireworks show off the ground. Later on I learned that an initial flare-up is part of most gas pressure lamp ignition routines. Fortunately, I also learned to

A Coleman can be a farmer's best friend during a power outage.

minimize the size of the fireball. But I continue to keep my head back and I still handle Colemans like Roman candles.

Safe Handling

There are two types of Coleman pressure lanterns—gasoline and kerosene. Both are fundamentally similar, incorporating hand-pressurized fuel tanks and one or two mantles. But each fuel demands its own distinct ignition procedure. The gasoline lantern offers the easiest start-up, but because of the highly flammable fuel it calls for the most cautious handling. The kerosene lantern is the trickiest to light, but the low flash point of the fuel makes it the safest. Illumination is slightly higher with the gasoline double-mantle lantern; the cost per hour is slightly lower for the single-mantle kerosene lanterns.

It is absolutely crucial to know what you're doing when you

operate a Coleman lantern. The combination of pressurized flammable fuel—especially gasoline—and a tricky ignition procedure that often involves a momentary burst of flame may cause injury or property damage if you do not follow instructions. The instruction pamphlets provided with each lantern all begin with boldface warnings, which can be condensed as follows.

- Do not use in unventilated areas.
- Use for lighting only.
- Never fill lantern or loosen or remove fuel cap while lantern is lighted, near flames or other ignition sources, or while top of lantern is hot to touch.
- Never allow any flammable material to come within two feet of the top and one foot of the side of the lantern.
- Replace any mantle with a hole in it.
- Never adjust light output with fuel valve or cleaning lever.
- Keep out of reach of children.
- In the case of gasoline lanterns, use only Coleman fuel or clean, fresh white gas. Never use fuel containing lubricating oils, lead compounds, or other metallic compounds. (Some unleaded automotive fuels may contain metallic compounds.) Coleman fuel is extremely flammable. Use the same care as when using gasoline.
- For kerosene lanterns, use only clean, fresh white kerosene.

Coleman makes several pressure lanterns: both double- and single-mantle gasoline burners and single-mantle kerosene lanterns. Their most popular lantern is the double mantle gas model. Its one-quart fuel capacity generates eight hours of 220 candlepower light, and the 3200 Btu per hour heat output is enough to take the chill off a tent or small room. The lantern stands fourteen inches high and weighs five pounds. Except for the mantles, the lantern comes pre-assembled.

Fueling

Whether you burn Coleman gas, white gas, or kerosene, it's a good idea to get in the habit of pouring the fuel through a filtered funnel. That will minimize the buildup of deposits in the generator, which leads to lower efficiency and eventual replacement of the generator. Before filling, close both the fuel valve and the pump knob firmly. During filling, the lantern should be in a level position. Tighten the fuel cap firmly after filling, and wipe up any spilled fuel.

Mantles

Unscrew the ball nut at the top of the lantern so you can remove the ventilator top and the glass globe. Then tie the mantles on with the strings tight around the grooves in the burner caps. Distribute the folds evenly. The flat side of each mantle should face the generator. Cut off surplus string. Always use Coleman mantles, which are designed for use with high-pressure fuel; LP tie-on mantles are larger, with a looser weave, and will not work properly. Now burn off the protective mantle coating by lighting—without touching the bottoms of the mantles. This should be done in a well-ventilated area, or outside, to clear the smoke. Be sure that the mantles are burned until only a white ash remains. Allow the mantles to cool before lighting the lantern. The mantles are very fragile and must not be touched. Now reassemble the lantern.

Pumping

With the fuel valve firmly closed, open the pump knob one turn. Closing off the hole in the knob with your thumb, pump approximately thirty-five strokes. (If the lantern is not full of fuel, more strokes are required.) Now close the knob firmly. Good air pressure is essential to lantern operation. In older lanterns, the leather in the pump seal may dry out and require oiling to hold air. To test the seal, close the pump knob firmly, place your thumb over the hole, and pump. If you feel little or no resistance, remove the pump, work several drops of oil into the leather, and replace the pump.

Lighting the Gas Coleman

Move the cleaning lever to the down (open) position. Insert a lighted match through the lighting hole at the base of the globe. Hold it near (without touching) the mantle, and now open the fuel valve ¼ turn only. The mantles will take a few seconds to light, during which there will be a burst of flame. When the mantles glow brightly —but with no flames—open the fuel valve fully. For good air pressure, pump up an additonal fifteen to twenty-five strokes, following the pumping method previously outlined. After the lantern has started, and whenever necessary during operation, rotate the cleaning lever several times to clean deposits from the generator, then leave it in the down position. Keep in mind that additonal pumping will be necessary at intervals during operation.

Extinguishing

Rapidly rotate the cleaning lever several times and leave in the down position. Close fuel valve firmly. The lantern will dim and go out in a half-minute or so.

Kerosene Lanterns

Kerosene-burning Colemans operate essentially the same as gas burners, with the exception of lighting. In the case of one model, a preheater cup is filled with alcohol (not gasoline or kerosene), which is match lit through the lighting hole. The flame then is allowed to consume nearly all the alcohol before the fuel valve is opened and the lantern lights. In the case of another, a squeeze bulb is provided with the lantern. Two bulbfulls of kerosene are squirted onto the wick material inside a pre-heater cup and screen. The kerosene must soak into the wick for one minute or so. The kerosene then is lit through the lighting hole, and the flame allowed to burn for ninety seconds. Then the fuel valve is opened and shut immediately. If the mantle burns brightly with no smoke, the valve can be opened fully. If smoke and soot billow up, extra preheat will be needed before opening the fuel valve.

Trouble Shooting

Flames other than at the mantle indicate flooding or a leak. Close the fuel valve firmly. Permit the flames to burn out and allow the lantern to cool. In the case of flooding, wipe up the fuel and start again. Otherwise, check for leaks and repair. If the lantern dims, pump in more air and give the cleaning lever a few spins. Leave the lever in the down position. If the lantern sputters or loses efficiency, or the mantles turn black, the generator may be clogged and need replacement. (Be sure to rinse the lantern fuel tank occasionally with fresh fuel to remove sediment, gum formations, and moisture accumulations; clean out the burners once a year.)

Oil the pump leather periodically to maintain the seal.

Indoor Gas Pressure Lamps

Coleman lanterns are intended primarily for camping and outdoor use. Two features of the lantern—its stormproof top and one-quart fuel tank—make it less than ideal for indoor illumination. The top blocks off overhead light, and the tank needs refueling every eight hours or so. However, a California lamp designer has modified the double-mantle Coleman for indoor use. Combining the mechanism of the Coleman lantern with a two-quart fuel tank, a cylindrical glass globe with open top, and a fiberglass or glass shade, the Wetmore-Ceres Company has developed a "Parlor Lamp." The lamp will burn Coleman gas or kerosene for up to eighteen hours per filling. The tank is pressurized with a separate air pump. It's the brightest indoor firelight available.

Two disadvantages of the traditional Coleman lantern are overcome in this Wetmore-Ceres Parlor Lamp. The two-quart reservoir doubles the fuel capacity of the Coleman, and the lamp housing has been opened up to yield more light. Designed for indoor use, it's popular with the Amish, who shun electricity. This modified Coleman also has a shade.

THE KEROSENE WICK LANTERN

It's not always necessary, or even desirable, to stoke up a Coleman for outdoor light. In some situations the Coleman is downright overpowering. You can't turn down a gas pressure Coleman—it's either on or off. And who wants a hissing, blinding beacon on the picnic table in the evening, when a quiet, flickering wick lantern will do? Or where a long-running, low-key, fuel-efficient light is all that's required? Besides that, mantle lanterns are poorly suited to rough handling. It's true that the lantern housing is sturdy and stormproof, but it doesn't take much of a jolt to break the mantle, and that means replacement because even a small hole in the mantle will tend to overheat and crack the globe.

Hence, the kerosene wick lantern, known to many people as the trainman's lantern or hurricane lamp, is preferred. It's hard to top for all-around outdoor light. It's not as bright as a Coleman,

The Dietz lantern.

but it's simple to operate, safe, and inexpensive—three qualities that go a long way in the country. Essentially, the wick lantern is an oil lamp inside a stormproof housing. But instead of removing the chimney to light it, the glass is lifted with a built-in lever. Otherwise, it is operated in the same manner as a kerosene wick lamp.

The Dietz Company of Syracuse has more than a century's experience in the kerosene lantern business. Today the Dietz lanterns are most likely the sturdiest available. The Dietz Air Pilot generates twelve candlepower for forty-five hours per filling. And if you're looking for light with an extra flair, in addition to clear globe glass, they offer red.

Liquid Propane Gaslight

It never fails to please me when visitors mistake our LPs for electric light. After nearly a decade of lighting with gas, it confirms what I've discovered: gaslight and electric light are a close match.

What is it about gaslight that puts it at the top of the list of non-electric lights? For one thing, a well-maintained gaslight delivers at least the equivalent of a 50-watt bulb, and that's minimum. I find it's usually closer to 75 or 100 watts. Just as important as brightness is the constancy of light. Kerosene or white gas mantle lamps can generate as much light as a gaslight, but not on a continual basis, and not without periodic tinkering. It's the steady feed of pressurized LP gas that makes for steady light. And there's no chance of a flare-up due to overheating.

In addition to its uniform quiet light, the gaslight is bright by design. With liquid propane gas piped to the light, there is no fuel reservoir forming the base as is customary on most firelights. That eliminates the characteristic shadow that falls below most lamps. Anyone familiar with Colemans, Aladdins, and wick lamps will appreciate this design superiority.

In addition, because of the steady fuel pressure and the permanent mounting, the delicate mantle is unlikely to break.

An Economical Solution

Right now in my neighborhood—central Vermont—liquid propane ranges between $.60 and $1.20 a gallon, depending on the dealer and the amount of gas consumed. (Ironically, those who conserve pay the most, and those who burn the most pay the least.)

A gaslight will run approximately forty-eight hours on a gallon of LP gas—that's twelve hours per pound. To figure out your hourly running cost, divide the price per gallon by forty-eight. My dealer charges $.97 a gallon, delivered. That means I'm paying just over $.02 an hour. If I burned an average fifteen hours of light a night (three lights, for instance, each burning five hours) it would cost me about $.30 a night, $9 a month, or about $110 a year.

It's interesting to compare running costs for electric and gaslight. The electric rate here is $.126 per kilowatt hour in winter and $.055 in summer. It's not really accurate to average the yearly rate at $.090 because lights peak in winter, but let's do it anyway, keeping in mind that the figures will be skewed a bit low. Say a 100-watt bulb will run for $.009 an hour. At first it appears that gaslight costs about twice as much as electric. But if you take into account the electric company's monthly surcharge—here it's $5.78 for a private household—think of how much extra gaslight you could buy every month without that extra fee: about six gallons of propane, worth 288 hours of light. That's close to three weeks of gaslight at the daily fifteen-hour average.

How about LP compared to kerosene, Coleman fuel, or white gas? I've found that none of these burns significantly brighter than LP, yet the fuel is often two to three times as expensive. Other factors in cost comparisons include the gaslight itself, tubing, and installation. Again, gaslight compares favorably with its closest indoor competition, the Aladdin. A single gaslight will cost about $30, plus the price of a few yards of copper tubing. If you ask your gas dealer to do the work, the total cost per light will be roughly $60, depending how many you equip, and what happens with inflation. However, it isn't difficult to install your own gaslights, and I'll explain the procedure later.

There are other ways to consider the economics of gaslight. Property tax usually includes an assessment for electrical hookups. Where I live, 200-amp service is valued at $150, 100-amp at $120. But there's no tax charge for having propane tanks at the house. Finally, there's the indirect value of gaslight. People are catching on to the bargain in land that's not served by electricity.

To ignite liquid propane, strike a match, hold it close under the mantle without touching it, and turn on the gas.

Safety

Fire and fumes are the two hazards usually associated with combustible lighting. For the most part these hazards are eliminated by gaslight design. According to the United States Consumer Protection Agency, most lamp fires occur when a burning light is upset. This is a real danger with table lamps, but not with LP gaslights, which are permanently mounted on the wall or ceiling.

Another frequent cause of fire is lamp flare-up. This happens when a fuel lamp overheats, usually during warm up or in a poorly ventilated space. This cannot happen with an LP gaslight because the rate of combustion is governed by tank pressure and the lamp valve.

Fires can also start when a lamp is set near a flammable material. But when installed properly, following the manufacturer's clearance recommendations, a gas lamp burns within its own safety zone, like a wood stove.

I've been asked if there isn't a chance that a burning gaslight might go out and fill the house with a cloud of explosive gas. I suppose it's possible. But nothing close to that has happened to me during more than eight years of gaslighting, and I've never heard of it elsewhere. I have heard of explosions in homes using gas that is piped in from a distant source. These result from disruptions in the supply line. Such an accident is highly unlikely with LP gas, because the household fuel supply is on site, under your control. It's even possible to shut off the gas tank outside when you're not using the lights or other appliances. This is a nuisance, and we don't do it. Instead, we follow one house rule— never leave a burning light unattended.

When fuels burn incompletely, as almost all fuels do to some extent, toxic carbon monoxide (CO) gas is produced. You can't see, taste, or smell CO, but it can make you sick or kill you. A gaslight is designed to burn at a fixed rate, with no adjustments in the level of illumination. Thus, by design, it is tuned to burn at peak combustion efficiency, minimizing the production of CO. Unfortunately, most gaslight manufacturers are reluctant to be specific beyond recommending adequate ventilation. One engineer at Coleman told me: "We'd lose our credibility if we started setting ventilation standards." You can take "adequate" ventilation to mean two things: enough air for the light to burn properly, and an air exchange rate generous enough to exhaust fumes.

Before the advent of tight houses, natural air leakage in a home usually provided enough ventilation. Today, the question of ventilation is more important. After talking with several gaslight manufacturers, and an importer of lighting accessories, I'd say that the consensus on gaslight ventilation rates hovers between two and three cubic feet of air per hour per light. In old (leaky) homes, natural air exchange should be more than enough for several gaslights. To ventilate airtight houses, a nearby window should be left open slightly at top and bottom. Everyone I talked to agreed that gaslights and small, airtight enclosures don't mix.

"We're trying to steer away from camper caps and ice shanties," one manufacturer explained to me. Said another, "I've never heard of anyone succumbing to gas fumes in a house or a cabin, but I have heard of it in an RV."

Location and Installation

The typical gaslight has four main working parts: the valve and wall bracket assembly, the valve cover, a glass globe, and a mantle. You can put a gaslight together in a couple of minutes, and, with the unit unconnected, move it around to judge how it will fit in various locations. If you haven't seen a gaslight working, check one out; it will make it easier for you to locate your light in a good spot. Remember: gaslights and tubing are permanently mounted, and changing locations can be troublesome and expensive.

There aren't many manufacturers of gaslights for indoor use. The Humphrey Opalite is by far the most popular gaslight in the United States, so it's the one I'll refer to. (The Falk gaslight is popular in Canada.) Humphrey has been shipping gas lamps from their Kalamazoo plant since 1901. They have a reliable product and a good spare parts network.

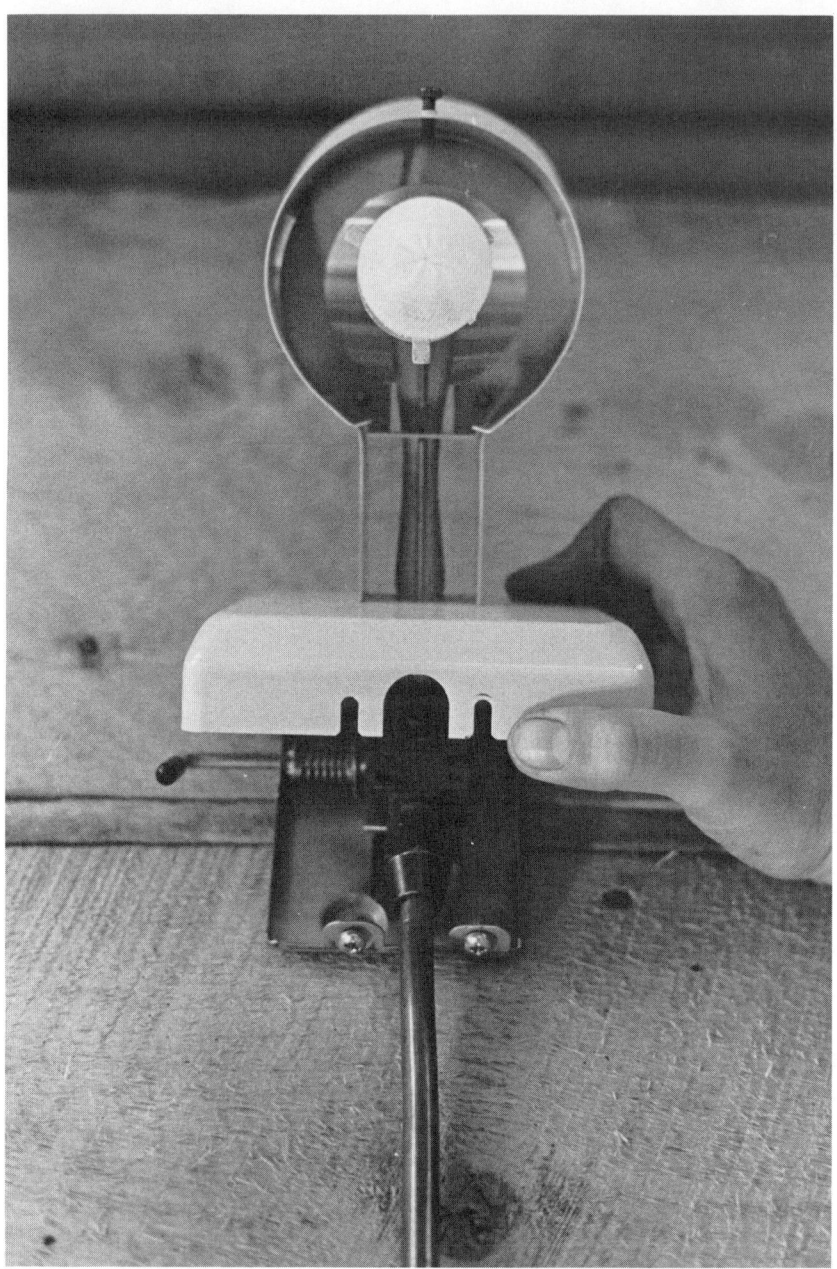

LP gas light anatomy. The wall bracket is the base of the unit and contains the threaded flare fitting for the upper supply line, the nozzle, and the on/off switch. The valve cover fits over the wall bracket and contains the bunsen, the burner nose, the mantle, the globe support, and an integral reflector that doubles as a heat shield.

Safety, illumination, and economy are the three main considerations when locating a lamp. The mantle is fragile—whether or not it's burning—and the glass globe is fragile too. So your lamp should be outside busy household traffic patterns, and away from work areas where percussion or vibration may cause damage (a workbench, for example). Liquid splashing on a hot gaslight can crack the globe, and a strong current of air can cause the light to pulsate, so keep locations a safe distance from kitchen sinks and cross-drafts.

Most important, follow or exceed the manufacturer's clearance specifications. One gaslight can generate 1,800 Btu every hour, and the temperature just above a light will be about 240°F. Humphrey recommends a clearance of four inches between a lamp and any combustion material, but I sited the light in my studio seven inches below a maple bookshelf, and I wouldn't get any closer. Allow at least a four-inch clearance on either side of the lamp.

At first glance a gaslight looks much like a traditional kerosene mantle lamp, but a closer look reveals a twist: the mantle is inverted. Upon ignition the direction of illumination is downward. Keep this in mind as you consider sites.

Note also that the Humphrey globe is divided into a northern hemisphere of frosted glass and clear glass below. Thus it is possible to locate lights at varying heights, depending on the objective (table or desk work, reading, general illumination). With the clear glass just below eye level, the frosted glass cuts glare. It's also possible to fit the all-frosted Falk globe on the Humphrey light.

To make room for the gas line there's a punch-out opening at the bottom of the valve cover. You'll want to be sure that a wall light mounts on a flat surface with no obstructions to the tubing. There is a fitting available that makes it possible to bury the gas line, but I don't like the idea. It makes alterations difficult, and who wants to tear out the wall to repair a leak?

Humphrey also makes a hanging assembly for ceiling suspended lighting, so that one, two, or three lights can be hung from an overhead position. This is especially helpful in kitchens, where a light over a work surface or table may be handy. The ceiling assembly includes a cover for an interior wall gas hookup, but again I'd recommend against it.

Another way to get around the limitations of a wall mount is to use a pole or post. Two out of three gaslights in my house are mounted on structual posts, and I sited them so that the backlight ordinarily lost on a wall doubles their effect. One is mounted on a post in the central room, with the forward light flow on the living area and the backlight on the dining table. The other is mounted on a post near the kitchen counter so that the light takes care of both the counter and part of the living room.

In an L-shaped room, mounting a light on one side of the inside corner adds roughly forty-five degrees of illumination. Even better, it's possible to devise a mounting base for the corner so that a maximum of backlight falls equally to the left and right.

Once your light sites are chosen, you can have your local LP dealer install the lamps, piping, and LP tank. If you have an LP gas tank setup already, he can simply splice into the tubing. A cutoff valve to the lights will enable you to repair or change lamps without shutting off all gas appliances. Be sure to ask if your dealer has experience installing gaslights. It's worth asking around to make sure you don't hire a novice.

If you don't have a gas tank already, the most popular location for one is usually out behind the house. Just make sure you'll have access to bring in the supply hose or fresh tanks. In northern climates, avoid sites where the tanks will be buried in snow. In very cold weather—below −32°F.—liquid propane ceases to vaporize, and the light dims.

You can also do the installation yourself. Two special tools are required: a pipe cutter and a flaring tool. Besides these, you'll need a screwdriver, an adjustable wrench, and a drill with a bit about 1/16-inch larger than the diameter of the tubing. Use malleable copper tubing: 1/4-inch diameter tube for runs of ten feet or less, 3/8-inch diameter tube for longer runs and wherever more than one light is served by a single run. When you buy the tubing, make sure it's plugged at both ends, and keep it that way until you are ready to install it, or you may wind up with moisture or spiders inside.

For each Humphrey light, these fittings are required, according to the tube size: either a 1/4-inch or a 3/8-inch flare male connector with 1/8-inch NPT straight fitting. If the light is to be connected through wall or ceiling, use a 1/8-inch NPT street elbow (male-female). The tubing is then flared and connected to the valve on the Opalite's wall bracket. Smear the joint with pipe dope before screwing it tight. At the other end of the tubing, your gas dealer probably will connect the tubing to the regulator. If you do it, use a flared fitting here as well. Although the tubing is malleable, carefully shape any curves in a tube run so they don't crimp, and use elbows for right-angle bends.

It's easiest to connect the tubing to the lamp's wall bracket valve before the wall bracket itself is screwed to the wall. The bracket's recessed screw holes hold the unit away from the wall so that air can circulate behind the gaslight, shielding the wall from high temperatures. Do not overtighten, and be sure not to twist the bracket. Now remove the globe and attach the valve cover to the bracket by hanging it on the two tabs at the top of the bracket. Swing down the valve cover so that the slots in the bottom engage the locking screws, and tighten them.

If you're already hooked up to the regulator, leave the lamp switch in the off position, and check your work for leaks. Gas dealers use a special sniffer sometimes, but I've come to rely on liquid soap. Simply pour some dish detergent on each connection, and look for bubbles. Your nose will also alert you. Added to the odorless LP is a scent to make the gas detectable. In a full tank it's faintly noticeable, but when it settles in a low tank it's hard to miss; it smells like a dead mouse.

Mantles: Tie-On and Preformed

Now the lamp is ready for the mantle. There are two types of mantles: tie-on and preformed. Each requires a different burner nose on the lamp. The tie-on is cheaper, but it's a little tricky to attach. The preformed mantle is easy to attach, but it's more delicate and expensive. I prefer the tie-on, although it may take you a few tries to get the hang of it. Because the tie-on is unbreakable before burn off, and slightly flexible after, it's recommended for RVs. By the way, don't use Coleman mantles on LP gaslights.

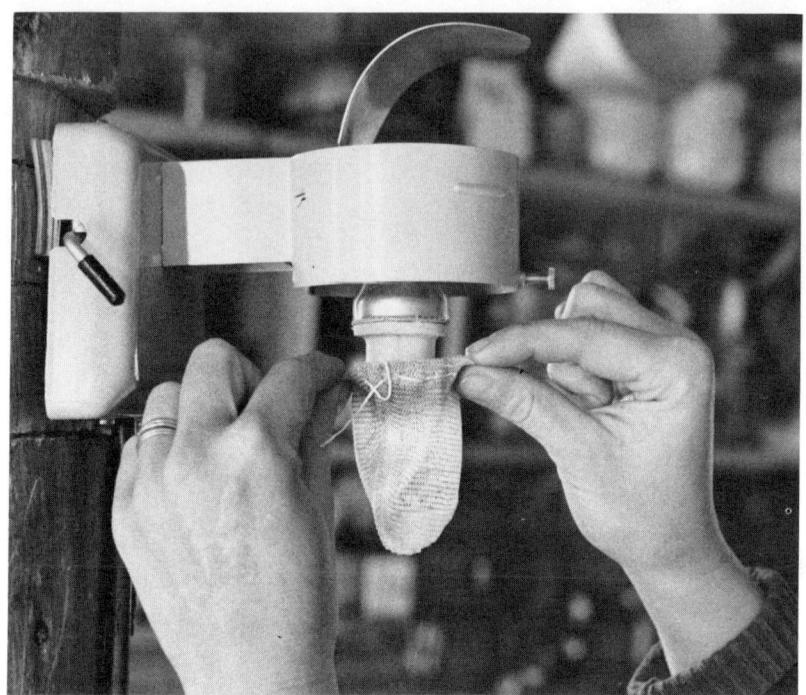

To install a tie-on mantle, tie the strings at the throat of the mantle over the ceramic burner nose and seat in the lower groove. The mantle must be distributed evenly around the burner nose. Pull the ends of the string so that the mantle and string are seated securely in the groove. Tie a double knot and clip off the excess string. After burn-off, it will shrink to its final form. Unlike the pre-formed mantle, the tie-on is manufactured in a flexible condition and is good for moving vehicles.

The pouch will not blossom properly with the lower LP gas pressure.

No matter which type of mantle you use, it's necessary to burn off the lacquer coating that protects the delicate fabric before use. The burn-off lasts several minutes, producing a smoky and perhaps unhealthy cloud that I don't want in the house. What I do is fire up the mantle outside while it's attached to the valve cover assembly. All you need to do is hold a lighted match just below the mantle until it ignites. Then back off and wait until the flame and smoke die out after the coating has burned. Now replace the valve cover assembly very gently on the wall bracket and tighten the screws. (If you don't handle the mantle tenderly,

The pre-formed mantle is simple to install: Lift the mantle by its porcelain base ring onto the burner nose and turn until it is seated. Do not touch the mantle! A pre-formed mantle requires a corresponding burner nose. Before use, the pre-formed mantle is more fragile than a tie-on.

it will break.) If you choose instead to preburn the mantle inside, do it with the windows open.

What you have after burn-off is a fragile, incandescent chemical ash that will glow in the heat of the burning gas vapor. (Initially, the tie-on will appear a bit misshapen after burn-off, but the first time the light is ignited it will shrink to its final form.)

Lighting

Turning the light on is easy. First, turn on the gas supply at the tank. Next, with the handle of the lamp in the off position, hold a lighted match just under the mantle and turn the gas on. You will have to depress the handle to release the safety lock. The first time you use the light, there may still be some air in the line, so you may need to light several matches. To extinguish the light, return the handle to the off position.

If you find that the light is too bright or produces glare, consider using a Falk gaslight or all-frosted globe, or improvise a shade. It's possible to wrap a curtain of aluminum foil around the globe and cinch it in place with a strand of wire; the foil can be lifted for brighter light. Or customize a fiberglass shade to wrap around the light.

Be careful not to overload your gas appliances. A single tank may not deliver sufficient pressure to power a house full of gaslights, stoves, water heaters, and so forth. If you see your lights fade when other gas appliances are burning, you may wish to hook the lights to a tank of their own.

Gaslight maintenance is no big chore: clean the glass globe occasionally and replace the mantle when it breaks. Check the burner nose periodically to make sure it's tight; the gasket tends to shrink with use.

I can still remember the first time I lit up a gas lamp. There was a muffled pop like a champagne cork, and a wave of light bathed the room. It was remarkably bright, with a warmth unmatched by electric lights. For a minute a low growl accompanied the glow as the lamp warmed up. Then the sound ebbed to the faint breeze of gas feeding the flame. It reminded me of the sound of a seashell at my ear.

Incandescent Mantles

Imagine the delight of a nineteenth-century lamp keeper upon discovering the newly invented Welsbach mantle. At that time fuel lamps burned at thirty candlepower, tops. Then along came a little thimble-shaped device suspended in the lamp flame that doubled or tripled illumination, and cut fuel consumption to boot—often at a rate between 300 and 400 percent. Who could resist? And indeed, after its patenting in 1885, the incandescent mantle beamed its way into millions of lives all over the planet, rescuing fuel lighting from the dark ages, saving eyesight, and conserving costs.

Radioactive Thorium

Nowadays, an estimated 25 million mantle lamps are in use annually in the United States alone, and a huge percentage of the world's population lights by fire. The mantle is little changed. Suspended at the center of our brightest non-electric lights—Aladdins, LPs, and gas pressure lanterns—it is still composed of 99 percent thorium and 1 percent cerium. As ever, this means the mantle emits radiation. You're not likely to find that disclosure alongside the superlatives in lamp ads, but it's no secret. Any decent encyclopedia will tell you that thorium is a naturally radioactive rare-earth metal—soft, lustrous, and silver white. It can be found abundantly in Brazil, India, Canada, and, in especially large deposits, New Hampshire. It is the most stable of the twelve known radioactive isotopes. During its decay, thorium emits both alpha and beta radioactive particles.

Incandescent mantles take various shapes to fit the lamps; they are not interchangeable. Clockwise from top left: the Aladdin mantle; a pre-formed Humphrey gas light mantle; a tie-on gas light mantle; and a tie-on mantle for a Coleman lantern.

What does this mean to the mantle user? Is there a health threat associated with mantle use? After decades of apparent indifference and ignorance (and perhaps intentional silence from mantle makers), lamp keepers have become aware of mantle radiation and are questioning safety. The centennnial of the Welsbach mantle seems a good time to look for an answer.

Before delving into the invisible realm of radioactive isotopes and millirems, it may be useful to have some background on the delicate network of metallic ash filaments we call the mantle. The mantle is the handiwork of chemist Carl Auer von Welsbach, born in Austria in 1858. *Scientific American* introduced its readers to his invention in 1887, reporting on a mantle demonstration at the Marlborough Gallery in London. Welsbach mantles attached to the gallery's ordinary gas fittings "emitted a white and brilliant light resembling somewhat of an incandescent electric lamp." It was the mantle's power to challenge electric light that excited people. Six years before von Welsbach's invention, Thomas Edison had introduced his incandescent electric light. Initially it appeared that Edison's new light would easily overshadow the country's fuel lamps. However, the mantle turned out to be even brighter than electric light. Eventually, of course, electric light grew brighter and moved into the majority of homes—at least in the industrialized world. Nevertheless, today the mantle lamp lives on in rural homes, vacation camps, boats, and recreational vehicles.

Dangerous Fallout?

I first became aware of a question of mantle safety in 1979. A rumor began to circulate among my neighbors that the smoke from mantle burn-off was dangerous, but it never amounted to more than a rumor. Certainly nothing on the mantle package hinted at trouble. In fact, the only caution associated with firelight came in the LP gas lamp installation instructions that warned against carbon monoxide fumes. But this, I knew, was a potential problem for all fuel lamps, with or without mantles, and proper ventilation was the straightforward solution. Nevertheless, as I had recently installed my first LP mantle lamps, I started to investigate.

Initially, I talked to Herb Rosenhagen, president of Humphrey Products, the manufacturer of my gaslights. I asked him about the rumors I'd heard that mantles might be toxic. He said that several years earlier he had read a newspaper story warning of the dangers of mantle burn-off.

"There may have been a problem," he explained, "but not with our mantles." He sent a follow-up letter detailing the mantle scare. It seemed that a student in California who had written his master's thesis on mantles had stated that the burn-off was harmful. His report was picked up by a newspaper and created some alarm. Rosenhagen decided to get a full analysis of the Humphrey mantles, which are manufactured by the Falk Company in England. Analysis was performed by Wilmar Associates of Houston, Texas, and issued in a report dated December 12, 1973.

The gist of the study is that a highly toxic metal, beryllium, is the constituent in question:

> "When incinerated there is no question that beryllium could be toxic. However, tests have been run on old mantles that did contain some beryllium in the hardening solution, and it was found that its content in the burned mantles was less than 50 ppm (parts per million). Under no circumstances could the same concentration be toxic.
>
> "Falks mantles do not have beryllium. Basically a mantle is made with thorium and cerium hydroxide impregnated in a viscose rayon fabric, but on soft mantles a hardening solution containing aluminum and magnesium is applied to the top of the mantle and the area of this application is indicated by the dyed section of the mantle. A spectrograph analysis of the ash on a burned mantle shows about 99.26 percent thorium oxide. The balance of material in the mantle is listed as follows: cerium, 1,000 ppm; aluminum, 3,000 ppm; magnesium, 3,000 ppm; sodium, 50 ppm; calcium, 200 ppm; titanium, 10 ppm; iron, 20 ppm; copper, 7 ppm; zinc, 5 ppm; silicon, 50 ppm; lead, 10 ppm; sulphur, 70 ppm; traces of zirconium, yttrium, gallium and boron. You can see from the above that there is absolutely nothing toxic in the burned or unburned mantles."

Finishing the report, I glanced up at the glowing mantle over my desk. It had been Welsbach's brilliant notion to cap a Bunsen burner with a fabric saturated in a chemical solution from which all organic matter could be burned off, leaving an incandescent

Mantles are coated with a protective lacquer that you must burn off before using, preferably outdoors.

metal ash. Some manufacturers hardened the mantles with heat-resistant beryllium, but the stuff proved toxic. Nevertheless, beryllium or no beryllium, the main mantle ingredient was 99 percent thorium—and it was then I discovered in my encyclopedia that thorium is radioactive.

Let me tell you about thorium. In 1828, a fellow named J.J. Berzelius gave the name to the oxide he had extracted from the rare mineral organite in Lovon, Norway. He named it after the highest deity of the heathen Scandinavians, Thor. Potent stuff. When heated, thorium ignites easily, burning with a bright white flame. It refuses to melt below about 1750 °C. Thorium is used in tungsten filaments for light bulbs and electronic tubes, welding rods, magnesium alloys, radiation therapy, and for the past hundred years, as the chief constituent in fuel lamp mantles.

Perhaps you can guess my reaction to this news—had I inadvertently planted three radioactive devices smack-dab in the middle of my house? I was shocked, to put it mildly. What was a rural lamp keeper to do with this information?

Certainly, the measure of mantle safety cannot be taken optically. To understand lamp safety, I had to slip into the invisible dimension of radioactivity.

My first contact was with Wilmar Associates, where the Humphrey mantles had been tested.

"People are too damn worried!" announced W.S. Saese, when I asked about the potential for trouble from thorium. "The activity is so low you'd have to swallow ten mantles to get an effect. The mantle is about as radioactive as television."

"Black and white or color?" I asked.

"Either."

Next I contacted Consumers Union, publishers of *Consumer Reports*. "We've never done a thing about gas lighting," I was informed. At underwriters Laboratory I drew another blank.

I tried the Vermont Health Department. Tom Neiman, liaison officer for various department offices, seemed surprised by my query. "It's the first question we've had about gaslight. We have no regulations regarding gaslight, and we have no authority in private homes. The regulations cover only rental and public buildings. In the rental housing code gas heaters must be vented to the outside. Proper ventilation is the main concern. In fact I have a part share in a cabin in the woods, and we use gaslights, and they work fine."

Marty Johansen of the Vermont Occupational Safety and Health Agency told me that his outfit had no pertinent data. "Back in the 1800s every house on the street had a gaslight," he said. "For many years gaslights have been going out of style. Now they're coming back, I suppose. I think a fire hazard might be a worry. I'd be concerned about pounding a hole in the wall and puncturing the pipe and getting a leak. As far as thorium goes, the amount is so small it's no problem."

I learned from the state Fire Prevention Office that any gas appliance must be approved by either Underwriter's Laboratory (UL) or the American Gas Association, and that gas dealers are supposed to install appliances in accordance with the National Fire Prevention Association code. But in fact, the state office has no jurisdiction over private homes.

My gas dealer confirmed that he goes by the NFPA code book specifications when installing appliances. Trouble is, he couldn't find gaslights mentioned in either the book or AGA publications. However, the latest Humphrey light is adorned with the stamp of approval of the Canadian Gas Association, providing correct installation procedures are followed. I asked John Doran of the CGA just what that meant.

"In general, gas appliances must be installed with an eighteen-inch clearance above and five to the side, according to the code. Also there are specifications for making sure installation is secure." CGA specifications also allow for gaslight valve leakage of no more than twenty cc an hour. "And that's very little." Doran emphasized.

I asked him about the mantle. "All I can say is that in 100 or so years there's not sufficient evidence of trouble. It's less radiation than you'd normally get," he concluded.

Next I turned to science departments at Dartmouth College.

"I don't think you have anything to worry about," was the response I got from Roger Soderberg, professor of chemistry. "Maybe from particles, but the pressure is low and decay is extraordinarily slow. The half-life of thorium is extremely long.

Gaslight is innocuous. It's been around 100 years and there have been no alarms."

Charles Drake, professor of earth sciences, agreed that thorium presented little threat. "The reason it's used is that it doesn't burn away, so nothing gets distributed. It's no more trouble than a rock sitting in a display cabinet."

At this point in my sleuthing I decided the mantle scare was a false alarm and dropped the matter. It wasn't until about three years later, in December 1982, that *The Mother Earth News* published an article called "The Hidden Danger of Lamp Mantles," by Mary Anderson. Her story reported on a class action suit against the Coleman Company and other mantle manufacturers. According to Anderson, a California health physicist named Walter Wagner had charged Coleman with endangering public health by using radioactive thorium in mantles. Specifically, Wagner sought to return $300 million in reputed damages to mantle purchasers and to require the company to begin using warning labels on their products.

Apparently, it was Wagner's contention that during thorium's decay, dangerous alpha and beta particles were emitted. The story painted a frightening picture: children, pregnant women, and unborn babies were at risk from "radium-laden fumes from lamp mantles" when a light was burning; and carrying a packaged mantle in a shirt or pants pocket exposed people to "significant" doses of beta radiation. "Users of mantle lamps should especially avoid breathing in the particulate ash or getting ash in their food," wrote the author.

Anderson confined her story essentially to a repetition of Wagner's allegations against the mantle manufacturers. However, *The Mother Earth News* editors wrote an accompanying piece detailing some experiments they'd conducted with a small radiation detector. Indeed, they found that lamp mantles emitted anywhere from 0.05 to 0.30 millirems per hour, depending on whether they were in the package or on a lamp. But they felt compelled to add that low-level radiation of this sort is emitted by watches with radium-painted dials and certain kinds of crockery and pottery, which actually emit more radiation than lamp mantles. Other sources of low-level radiation are old clocks and timepieces, household smoke detectors, old kitchen timers, record and film static eliminators, concrete, brick, and in fact, rocks and soil—good old mother earth.

"Perhaps the main point of MOM's little test," the editors summed up, "is that it awakened us to the prevalence of sources of low-level radiation.... The big question, of course, is what dosage of low-level radiation places a human being at risk. This may never adequately be answered (and there may well be no 'safe' dosage)."

Nowhere was it mentioned that the United States Code of Federal Regulations classifies thorium under the general heading "Unimportant quantities of source material" from naturally radioactive metallic elements, and exempts from all licensing "any quantities of thorium contained in (i) incandescent gas mantles, (ii) vacuum tubes, (iii) welding rods, (iv) electric lamps for illuminating purposes."

Whom to trust? I decided to get in touch with Coleman. Three years before I had asked Wilbur Townsend, technical research engineer for Coleman, if mantles posed a health threat. Not a chance, he told me. "As far as thorium is concerned, we have people in the mantle shop continuously exposed, some of them for forty years. They undergo periodic whole-body checkups in Denver. And they come out clean as a hound's tooth."

Now, in the light of Wagner's lawsuit, I wondered how the Coleman Company would respond. In February 1983, I received a letter from Harold J. Pfountz, a Coleman corporate attorney, which detailed the highlights of their defense. "We at the Coleman Company," he wrote, "are firmly convinced that the incandescent gaslight mantles which we offer are safe. This position has been confirmed by recent studies performed both by the Nuclear Regulatory Commission and by independent radiation health physicists on behalf of the Coleman Company, Inc." Enclosed with the letter were copies of several documents substantiating in detail the Coleman position. Two of the affidavits were prepared by Dr. Robley D. Evans, a radiation health physicist with impressive credentials.

Dr. Evans's qualifications include fifty years of direct experience in the measurement of radioactivity and ionizing radiation and in the evaluation of biological and medical effects of radiation. His first experimental studies of radioactivity levels in a thorium gas mantle manufacturing plant were published in 1938. He is a charter Certified Health Physicist of the American Board of Health Physics, a former national president of Radiation Research Society and the Health Physics Society, an alumnus of the California Institute of Technology (B.S., M.S., Ph.D.) and now professor of physics, emeritus, at the Massachusetts Institute of Technology, where he served on the faculty from 1934 to 1972 and as director of its radioactivity center and consultant to its medical department.

In his declaration to the court, supplementing his own data, Dr. Evans occasionally refers to the United States Nuclear Regulatory Commission's 1981 document NUREG/CR-1910, "An Assessment of Radiation Doses from Incandescent Gas Mantles that Contain Thorium." This is a ninety-two page, five-year study that considers limits of exposure for people involved in the manufacture, distribution, warehousing, sale, use, and disposal of thorium gas mantles. The gist of the report is that whether one is a thorium mantle user or not makes no significant difference. Dr. Evans states he had had "previous occasions to study technical documents of the NRC and I have been publicly critical of several. In the present case the document represents, in my view, a scientifically competent, thorough, professional-level investigation, and I agree with their methods and their conclusions." (Wagner did not refer to this document in his suit and may not have been aware of its existence.)

Statistics

The following are some of the highlights of Dr. Evans's declaration:

> "The naturally occurring radionuclides are universally distributed in all materials. In the case of rocks, soil, and dirt, the average U.S. terrain contains, per square mile and to a depth of one foot, some 2 grams of radium, 6 tons of uranium, and 20 tons of thorium. In one thorium gas mantle, the amount of thorium (300 milligrams) is substantially the same as in 70 pounds or in about one cubic foot of ordinary rock, soil, or dirt."

"In general, average doses to mantle users are about 1/1000th of the natural background radiation in which we all live."

It is the possibility of volatilization and release of some or all of the radioactive decay products of the mantle that had been Wagner's main concern. He states that at each lighting of the mantle dangerous amounts of "radioactive materials" will be volatilized, vaporized, or otherwise emitted into the surrounding air. Yet, according to Dr. Evans, the actual operating temperature of thorium mantles is at present unknown, making it impossible to determine those emissions. Moreover, in its study of thorium mantles, the Nuclear Regulatory Commission stated that even in a "worst-case" scenario, where all decay products of the thorium series were released (a virtual impossibility), the effect would be innocuous.

Wagner states, "there is very little hazard associated with the other components other than radium in the mantles." Dr. Evans counters that the intake of radium that has been found by direct observations to be innocuous to humans corresponds to the radium content of 8,000 to 10,000 mantles.

Studies of people involved in extracting thorium and manufacturing mantles, who would have much greater exposure to all the radionuclides in the thorium than mantle users, "support the conclusion that no health effects are to be expected in users of thorium mantles," according to Dr. Evans.

From my reading of Dr. Evans's report and several excerpts from United States government studies, it seems apparent to me that Mr. Wagner will have a tough time convincing many people of his charges. Indeed, in January 1983, the Nuclear Regulatory Commission filed a brief in connection with the suit recommending that the court dismiss Wagner's claim seeking a labeling requirement, as well as all other claims in the complaint that come under the commission's jurisdiction. As of September 1984, Coleman was waiting for the court's verdict.

Guidelines

I'm grateful to Walter Wagner for bringing this subject into the open. It's made me aware of the potential for trouble, even if it is minute. And so, in my house, I'll follow some of Wagner's suggestions for mantle safety:

- Mantles should not be handled by children, or pregnant or nursing women.
- The initial lighting should be done in a well-ventilated place.
- Mantles should never be placed in pockets.
- Users should avoid breathing the ash and avoid allowing the ash to contaminate food.
- Mantle lamps should never be burned without the chimney in place.
- Used mantles should be disposed of in a tin can with a tight-fitting lid in the local landfill or dump.

These seem to be simple enough precautions to add to the traditional lamp keeper's caveats regarding ventilation and fire.

The Return of Firelight

Over the past few years there's been a revolution in wood and coal stove design; after decades of blackout, firelight is being restored to heating stoves. Crimson flames flicker behind glass door panes and stove-length windows, and the pages of magazine ads and brochures are ablaze. I'm looking at one ad now: a cozy couple snuggling up in front of their glazed firebox. NO COMMERCIAL INTERRUPTIONS, the headline crows. BRIGHT COLORS, MESMERIZING ACTION. Better than a Sony, indeed.

Over the centuries the trend in wood heat had been clear: the hearth warms up in inverse proportion to the light. Fireplaces waste heat—so close them up in a metal box. Mica and isinglass windows? Too fragile and not truly transparent. With the coming of the airtight stove, firelight was totally eclipsed.

Following the hit sales of the mid-seventies, the stove business cooled off under a drift of unsold black boxes. Then the wood stove caught up with the space age. High-temperature, stress-resistant ceramic glass, developed for missile nose cones, drew the attention of the stove designers. In 1979, responding to requests, Corning began to market a newly developed transparent stove glass called Pyroceram, and similar (and sometimes better) ceramic glasses started arriving from West Germany (notably Robax glass, made by Schott) and Japan (Nippon Electric Glass's Neoceram).

The Merits of Glass

The primary advantage of a fire view is obvious: the window offers an instant reading of the condition of the fire, easing reloading—no damper and door to open to check the fire, no smoke spilling out into the house. And after dark the window doubles as a night-light.

Stovekeepers will recognize the practical merits of a fire gauge and night-light, but the Btu's that go with firelight may come as a surprise. While steel or iron stoves launch about 90 percent of their heat upward, a glass pane beams more outward. Glass doors affect the way radiation heat is transmitted. Pass your hand across the front of a glazed stove and feel the heat increase over the window; it's in the nature of glass to beat metal at transmitting heat. A pane of glass contains less mass than a similar one of steel or iron. Invisible electromagnetic waves in the infrared band of the spectrum also radiate from the flames and red coals, passing through the glass to heat on contact. Altogether, clear glass transmits up to 25 percent more heat than an equivalent expanse of metal.

Moreover, there's the psychological wave length. It's been proven that color perception can alter mood and biochemical reactions. Food looks brighter if you're hungry, and a dark room seems cold. But flood the scene with stovelight, and you might warm up more than your toes.

The fire-view stove appears to be a superior investment too. As one stove designer told me, "It's the fire view that sells the stove." The stove buyer who foresees a resale or trade-in should consider one. At least that's what I thought when the time came for me to choose a wood stove for my new studio. Besides, I could use the extra candlepower.

Three Fire-Views

So I began visiting stove dealers and writing letters to builders. What's the safest, warmest, brightest, glazed stove, I asked. I classified them into three types.

One type of stove is designed primarily for a fire view. Usually it's a simple steel unit with a large window on the side. The glass is boxed into an offset frame something like a window box, full of flames instead of flowers. There's often a sill of about six inches between the glass and the stove wall opening. This

Franklin-type glass door stoves operate open or closed, both with a fire view.

type of stove is *not* airtight. The glass sits slightly loose in its molding, and supplementary vents draw air into the window box. This flow of air helps keep the glass cool and soot-free. Although these stoves are not technically airtight, they are still much more efficient than an open hearth. On some of these stoves the glass window can be closed off to tighten up the stove—but there goes the view. Stoves made from thin steel have problems of poor heat retention.

The second type of glazed stove is the airtight box stove—steel and/or iron—with glass in the front loading door. It generates higher temperatures than the vented window box stove and requires better thermal- and stress-resistant glass. Glazing an airtight like this wasn't really feasible until recently. In fact, less than a decade ago one designer had to use nine-inch Pyrex pie plates for windows. Pyrex was all right to a point—about 450°F. Beyond "broil," alas, the glass had a tendency to explode. Now the same stove maker uses "rocket" glass, heat resistant up to 2000°F. The lingering problem with the fire-view box stove is that the logs are loaded from the front, and the door closes against the ends of the logs. As one designer wrote me, "occasionally breakage results when the glass, which is on the loading door, is used to push logs into the stove, or strikes a log that is too long for the stove."

The third type of fire-view is the airtight Franklin-type stove with glass in the fireplace doors. Ben Franklin had intended to combine the efficiency of the closed metal stove with the "healthful and cheering" effect of the flames. Trouble was, with the doors open the stove threw most of its heat up the chimney. It wasn't until stove makers put glass on the Franklin that you could have the views and heat both. Another advantage of the fireplace layout is that the logs lie parallel to the stove front, reducing chances of logs breaking the glass. Also, because the logs are viewed broadside, they yield more light. These stoves are usually cast iron, ceramic, or soapstone, with superior heat mass.

All things considered, it seemed to me that the airtight glazed Franklin offered the most. For unobstructed viewing, the stove could be run with doors wide open. To amplify heat and save fuel, the doors could be shut and the fire viewed through the glass. I knew that a slew of stove builders were producing the type, but my choice was easy. Not far north from where I live is the home of Vermont Castings, pioneers in the design of fireplace stoves. They offer several models, all variations on one theme: highly baffled, cast iron airtights with swing-open viewing doors. The Defiant and the Vigilant, their first models, appeared with solid iron doors. Next was the Resolute: compact enough for my studio, and there in the front door was the window I wanted.

I drove up to Randolph for an afternoon of grilling the stove makers. I talked to people in customer service, design, and testing. Customer requests had persuaded the designers to try glass doors; I learned that early Vermont Castings stoves incorporated a Japanese glass, which is clearer than Corning but occasionally crazes, covering over with a network of tiny cracks that blur the view. An ongoing problem with all of the glass-door stoves is carbon residue on the glass. For coal burners, this is not an issue; high temperature and low carbon content keep the glass clear. But wood burners had a gripe—foggy glass. Most stove manufacturers recommend running the stoves hot to avoid this problem.

Breakage turned out to be less trouble. After winnowing out some inferior views, the designers selected Robax, the ceramic pane made by Schott, a West German glassmaker. In addition to being heat-resistant, Robax (like the Nippon and Corning products) is strong enough to pass the stove industry impact standard: a one-pound ball dropped on the glass from a couple of feet. I heard you could even throw snow on a hot window without cracking it. Besides having inherent resistance to impact and thermal shock, the Resolute stove glass is protected by an exterior iron grill.

In the testing shed I was escorted alongside a row of coal burning stoves. The windows were as bright as orange traffic lights. Then I spotted a stove with pitch-black glass. A wood burner.

"Is it always black?" I asked.

A stove tester opened the damper and added a scrap of dry lumber. The soot burned off like a rising curtain, releasing a flood of crimson. I was sold.

A blizzard blew in that night. By the time I had the stove ordered and a neighbor and his pickup corraled, three feet of snow covered the hills. We had to push the 300-pound stove crate up the road on a borrowed dogsled. My pal rode the sled back down, and I began to unpack. I followed the assembly instructions with a single exception. The glass is designed to be installed with an airtight gasket. I left it off, having received an unofficial tip that a loose window stays cleaner. Apparently the extra air circulation flushes away soot from the glass. I noticed that the window rattled a bit whenever I opened the fireplace door.

I situated the stove in the southwest corner of my studio, facing my desk, so that while I work, as I do now, I can look across the room into the fire. That first evening I watched the glass fill up with flames. Orange hues dappled the spruce walls. It was a whole new way to keep warm.

But after a couple of fires I thought the honeymoon might be over. Lighting the stove one morning, I saw that the window had smudged over. Perhaps it was that green maple I threw on? I loaded some bone-dry beech and kindled a wide-open fire.

But I couldn't duplicate the disappearing act I'd seen in Randolph. Some carbon residues burned off, but a dark stain remained. It took a winter of dedicated pyromania to erase that smudge. What follows are the highlights of that stove study.

Grooming the Glass

First, I found that it's possible to regain clear glass using the window cleaner provided with the Resolute. It's called Fire-Brite. If you want to hang onto your fingers, wear gloves during application. (Fire-Brite is full of sodium hydroxide.) Usually it takes two treatments at twenty-minute intervals to clean the glass. Oven cleaner also works well.

Better yet, use a natural cleanser provided by your stove. Dampen a wad of newspaper, dip it in the ash bucket, and give the glass a polish. It works splendidly, and you can't beat the price.

It helps to crack the fireplace door a bit while waiting for the fire to catch on. That speeds up ignition and prevents smudging.

It's important to burn dry wood for a hotter fire with less accumulated carbon on the glass. Friends with coal stoves have encountered even fewer problems with smudging.

Don't add a lot of wood at once—it cools the fire down.

Run the stove with the damper open. This keeps the fire burning hot and allows residues to exhaust. It's less efficient, of course, and you won't do it all the time, but there's no getting around it: with the damper closed the glass is likely to blacken.

Keep the chimney clean. The better the draft, the cleaner the glass. In fact, my hunch is that if I jumped from six-inch to eight-inch stovepipe, the view would improve.

Eventually I installed the glass using the airtight gasket, and the stove actually burned more efficiently without affecting the view (and now the window doesn't rattle).

One day quite by chance I discovered my favorite glass cleaning trick. I had just lit the stove. The glass was smudged black, and I didn't have time to rub it down with the cleaning solution. It happens that I had just cleaned my chain saw file with a wire brush. I picked up the brush and started to sweep the glass.

The carbon stain peeled right off! I've heard that scraping can damage some stove glass, but on the Resolute it doesn't leave a scratch. (For the record, the people at Vermont Castings do not recommend this method.)

Is all the elbow grease and extra wood worth it? I'm not sure you can measure firelight and come out in the black. But that's not the whole point. As a friend of mine said simply, "You do it for the glow."

STOVE GLASS

There are several types of heat-resistant glass suitable for stoves: tempered soda lime glass, tempered borosilicate glass, silica glass, and ceramic glass. Tempered soda lime glass is very clear but the least heat-resistant. It's suitable only for fireplaces or stoves with window vents. Should this type overheat and break, glass fragments may scatter explosively.

Tempered borosilicate glass, such as Corning's Pyrex, works at intermediate temperatures. This glass is used frequently in ovenware, panel heaters, and laboratory glassware. It will handle almost any thermal shock that may be encountered at operating temperatures below 450°F.

Silica glass, such as Corning's Vycor, is designed for ultra-high-temperature performance. It will handle any anticipated temperature and thermal shock requirements. It is suitable for both wood and coal burning stoves. Vycor is heavily rippled and not very clear.

Ceramic glass is composed of manmade materials. Its thermal properties are somewhat below ultra-high-temperature silica glass, but still suitable for most wood and coal stoves. One of its advantages is its low thermal expansion; it can be used over large expanses. And it can be remarkably clear, depending on the manufacturer. Corning's Pyroceram is tinted amber and looks out of focus; Nippon Electric Glass's Neoceram is sharper and untinted. Schott's Robax is the clearest.

Sources

Candles

Hamm's Bee Farm
Route One
Mason, Wisconsin 54856

Beeswax candles made from capping wax; twelve-inch and seven-inch. Discounts for quantity orders. Hamm's will make candles from your own beeswax for $3 a pound; one pound produces six twelve-inch candles (twenty pound minimum).

Eddie Bauer
P.O. Box 3700
Seattle, Washington 96124

Brookstone
127 Vose Farm Road
Peterborough, New Hampshire 03458

Fifty-hour microcrystalline emergency candles.

Gardener's Eden
P.O. Box 7307
San Francisco, CA 94120-7307

Citronella insect-repellant candles.

Kerosene Wick Lamps

Cumberland General Store & Catalog
Route 3
Crossville, Tennessee 38555

Extensive selection of kerosene lamps, chimneys, burners, wicks, brackets, shades; Aladdin lamps and parts. Catalog $3.

Timeless Tools
P.O. Box 36
Deerton, Michigan 49822
Kerosene lamps and parts; Aladdin lamps and parts.

Gatco, Inc.
1420 Carroll Avenue
San Francisco, California 94124
Wholesale supplier of Duplex brass burner kerosene lamps and other fine lamps. Write for location of your nearest dealer.

The Washington Copper Works
Washington, Connecticut 06793
Hand-wrought lighting fixtures, including a "lantern heater." Catalog $2.

Faire Harbour, Ltd.
44 Captain Pierce Road
Scituate, Massachusetts 02066
Parts and 1-K kerosene for wick lamps; Aladdin lamps and parts. Write for catalog.

South Road Pottery
South Road
Bradford, Vermont 05033
Duplex brass burners and ceramic kerosene lamps.

The Mantle Lamp Supply Company
P.O. Box 959
Sutton West, Ontario, Canada L0E 1R0
Kerosene lamps, parts, and shades. Lighting books. Reproduction parts for collectors. Aladdins. Catalog $1.

W.N. de Sherbinin Products, Inc.
Hawleyville, Connecticut 06440
Kerosene lamps, burners, chimneys, shades, and parts. Wholesale supplier. Write for location of nearest dealer.

Lehman Hardware and Appliances
P.O. Box 41
Kidron, Ohio 44636
Lamps and parts.

Nowell's, Inc.
Box 164
Sausalito, California 94965

Lamps and parts.

Don Snyder Imports & Antiques
Box 207
Comptche, California 95427

Kosmos kerosene round wick lamps.

Aladdin Lamps

Aladdin Industries
P.O. Box 100255
Nashville, Tennessee 37210

Aladdin lamps and parts.

LP (Liquid Propane) Gas Lights—Indoor

Humphrey Products
P.O. Box 2008
Kalamazoo, Michigan 49003

Single wall model; ceiling mounts for two or three lights; tie-on and preformed mantles.

Coleman Company
250 North Saint Francis Avenue
Wichita, Kansas 67202

Table-model cartridge LP lanterns adaptable to bulk tanks.

B.D. Wait Company, Ltd.
430 Wyecroft Road
Oakville, Ontario, Canada L6K 2G9

and

590 Hodge Street
Montreal, Quebec, Canada H4N 2A4

Falk gaslights; single and double wall models; double overhead. Parts. Frosted globes.

Superior Propane
1865 Leslie Street
Don Mills, Ontario, Canada M3B 2M4

Falk and Humphrey gaslights, parts, mantles, accessories, and service. Canada-wide mail order distribution. Local distributors in all provinces except Saskatchewan, Prince Edward Island, and the Territories. Write for catalog, prices, and address of nearest outlet.

ICG Canadian Propane, Ltd.
9765 63rd Avenue
Edmonton, Alberta, Canada T6E 0J7

Falks, Humphrey, and Primus lighting fixtures, parts, accessories and service. Outlets in all provinces and territories except maritimes. In Quebec, ICG is called Gasbec. Write for complete catalog, price list, and address of nearest distributor.

Phillips Lamp Shades, Ltd.
172 Main Street
Toronto, Ontario, Canada M4E 2W1

Humphrey gaslights, Aladdins, and Optimus pressurized kerosene lanterns and stoves. Write for brochure on all products and prices. Service and parts for all Aladdin products.

LP (Liquid Propane) Gas Lights—Outdoor

Coleman Company
250 North Saint Francis Avenue
Wichita, Kansas 67202

Table and yard lights fueled by LP cartridges or bulk tanks.

Lanterns

R.E. Dietz Company
225 Wilkinson Street
Syracuse, New York 13204

Kerosene wick lanterns. Special solid-brass models available.

Coleman Company
250 North Saint Francis Avenue
Wichita, Kansas 67202

Kerosene and gas pressure lanterns, propane lanterns, mantles.

Mantles

John D. Roba Company
P.O. Box 392
Swormville, New York 14146

LP gaslight and Coleman mantles.

Kerosene and Gas Pressure Lamps—Indoor

Wetmore-Ceres Company
998 30th Street
Richmond, California 94804

Homestead parlor lamps and parts.

Books

Oil Lamps: The Kerosene Era In America
by Catherine M.V. Thuro
Wallace-Homestead Company
1912 Grand Avenue
Des Moines, Iowa 50305

Lamp collectors "Bible."

Aladdin: The Magic Name In Lamps
by J.W. Courter
Wallace-Homestead Company (address above). Comprehensive history.

Kerosene

M & M Chemical Sales
Monson, Massachusetts 10157

1-K kerosene (TruLite).

FIREVIEW WOOD AND COAL STOVES AND FIREPLACE INSERTS

The Fire-View (M 270 SGR)

Window	10″ × 20″ Clear soda lime tempered glass
Material	Steel
Fuel	Wood
Height	28½″
Width	23½″ (28½″ with optional blower)
Depth	28″
Weight	238 lbs. (with firebrick)
Flue diameter	7″
Log length	24″

The Fire-View was introduced in 1969, the first of the new breed of glazed stoves. The stove is not airtight because the offset window box draws in air to keep the glass cool. The window can be covered for unattended burning. There are two flat cooking surfaces on top. Coal-burning models available.

Fire-View Products, Inc.
P.O. Box 370
Rogue River, Oregon 97537

The Russo Glass-View, High-Heat Stove

Window	11″ × 14″ Corning Pyrex
Material	Steel
Fuel	Wood
Height	27″
Width	30″
Depth	22½″ (32½″ with optional electric blower)
Weight	265 lbs.
Flue diameter	6″
Log length	24″

Another of the non-airtight, offset boxed window stoves, with glass easily removable for cleaning. Pickets protect the glass from shifting wood. Natural convection heating from internal ducts, with optional electric blower available. A door behind the glass can be closed for longer airtight burning. Coal and coal-wood stoves and fireplace inserts available.

Russo Manufacturing Corporation
87 Warren Street
Randolph, Massachusetts 02368

The Elm

Window	9¾" diameter circular; double layers of glass, Pyrex, and Pyroceram.
Material	Steel body, cast-iron front
Fuel	Wood
Height	25½"
Width	23"
Depth	Three sizes available: 50", 38", 32"
Weight	375, 280, 235 lbs.
Flue diameter	8"
Log length	36", 24", 18"

A veteran glazed airtight with distinctive tree insignia in front of the window. The double-plated window includes Pyrex on the exterior and Pyroceram inside. Vermont Iron Works now offers an Elm with a catalytic combustion device in the flue to clean up emissions.

Vermont Iron Works
Warren, Vermont 05674

The Quaker Box Stove (Buck II)

Window	5" × 8" Corning Vycor
Material	Steel body, cast-iron front
Fuel	Wood
Height	30¾"
Width	16¾"
Depth	35"
Weight	450 lbs.
Flue diameter	6"
Log length	28"

88 ALTERNATIVE LIGHT STYLES

Another front-loading glazed-door stove. The flat top is handy for cooking. The window is encircled by the antlers of a bronze deer. Available in three sizes.

The Quaker Stove Company
Kumry Road
Trumbauersville, PA 19970

The Moravian Parlor Stove

Window	Double doors, each 5¾" × 10¼"; Corning Vycor
Material	Steel body, cast-iron front
Fuel	Wood
Height	36½"
Width	29"
Depth	24½"
Weight	460 lbs.
Flue diameter	8"
Log length	25"

This Franklin-type stove allows open- or closed-door burning. Fireplace insert available.

The Quaker Stove Company
Kumry Road
Trumbauersville, PA 19970

The Vigilant

Window	Double doors, each 6½" × 8¼"; Robax ceramic glass.
Material	Cast-iron
Fuel	Wood or coal
Height	32"
Width	28¾"
Depth	19½"
Weight	295 lbs.
Flue diameter	8"
Log length	20"

The Resolute

Window	Two panes butted together making one arched window roughly 6" × 11"; Robax ceramic glass.
Material	Cast-iron
Fuel	Wood or coal
Height	26"
Width	26½"
Depth	17"
Weight	253 lbs. wood; 306 lbs. coal
Flue diameter	6"
Log length	16"

Lots of options in these stoves: open or closed door fireview, wood or coal, front or top loading, wide-open or baffled burning. When the damper is closed, heat circulates through the longest flame paths of any stove available. Cast-iron cooking griddle on top.

The Intrepid

Window	Two panes on double doors. Roughly 6" × 15," Robax ceramic glass
Material	Cast-iron and firebrick
Fuel	Wood or coal
Height	25"
Width	21¼"
Depth	18"
Weight	200 lbs. wood; 225 lbs. coal
Flue diameter	6"
Log length	16"

The Vigilant, The Resolute, and The Intrepid from:

Vermont Castings
Prince Street
Randolph, Vermont 05060

Hearthstone II

Window	Double door window panes roughly 6" × 12" total
Material	Soapstone
Fuel	Wood or coal
Height	26"
Width	27"
Depth	21"
Weight	475 lbs.
Flue diameter	6"
Log length	21"

The heat-retaining quality of soapstone sets this stove ahead of others in overnight heating power. Choice of gray or green soapstone. Hearthstone also makes a bigger wood-burning model.

Hearthstone
RFD #1
Morrisville, Vermont 05661

The Leyden Hearth Fireplace Insert

Window	10¾" × 15," ceramic glass
Material	Steel and firebrick
Fuel	Wood
Height	26"
Depth	26"
Weight	250 lbs.
Flue diameter	6" × 9" with adapter for 8" pipe optional
Log length	24"

Flexible panels allow the Leyden insert to fit into almost any fireplace, including shallow hearths. Internal ducts produce natural convection heating. An optional electric blower increases Btu output 20 percent. The largest glazed fireplace on the market.

Leyden Energy Conservation Corp.
Brattleboro Road
Leyden, Massachusetts 01337

The Kroupa Stove

Window	9″ × 13″ Corning Vycor
Material	Steel
Fuel	Wood
Height	Cooking surface 32″; oven top 48″
Width	46″
Depth	22″
Weight	600 lbs.
Flue diameter	8″
Log length	16″

Here's a stove for heating, cooking, and illumination. Its unique design incorporates an oven (13″ × 13″ × 20″); a cooking surface (22″ × 41″); and two fire levels (high grate for quick heat, good for summer cooking). There are glass panes on both ends of the oven and on the firebox door. A reflector plate can be attached to the firebox glass to keep in heat during oven cooking. The reflector plate doubles as a window cleaner, radiating heat onto the glass to burn off soot and creosote.

The Kroupa Stove Company
R.R. 2
Oliver, British Columbia
Canada V0H 1T0

The Kent Tile Fire

Window	10″ × 12″ Pyroceram
Material	Steel
Fuel	Wood
Height	$22^{7}/_{16}$″
Width	$20^{7}/_{8}$″
Depth	$24^{3}/_{8}$″
Weight	220 lbs.
Flue diameter	6″
Log length	18″

The Kent Tile Fire is a free-standing stove with a unique feature: the tile sides are interchangeable—select your favorite design. Air spaces between the tiles and the stove wall draw up cool air, circulate it over the firebox, and return it to the room as hot air. The Kent's twice-burning action offers heating efficiency and an exceptionally clear fire view.

Kent
59 Tidal Road
Mangere
Auckland, New Zealand

In the United States, call 1-800-457-4577 for location of distributors (in Oregon 295-0121).

MUSEUMS

The Winchester Center Kerosene Lamp Museum
100 Old Waterbury Turnpike
Winchester Center, Connecticut 06094
(203 379-2612)

Five hundred Kerosene lamps dating from 1856 are on display in this former country store overlooking a small village green in northwestern Connecticut. Lamps and lamp parts. Antique catalogs.

Index

A

Aladdin lamps, 22–36
  comparative fuel costs, 50
  incandescent mantle for, 62
  source for, 83
Aladdin paraffin oil, 32
Alcohol torch, for Coleman lanterns, 40
Argand, Aime, 19, 23

B

Beeswax, 7–8
Beryllium, in incandescent mantles, 64, 66
Bobeche, 10
Books, sources, 85
Bunsen, for LP gaslight, 54
Burners
  for Aladdin lamps, 25, 27–28
  for kerosene wick lamps, 16–17

C

Canadian Gas Association, 67
Candela, 7
Candles, 5–10
  making, 5, 7, 8
  sources for, 81
Carbon monoxide, 52
Cerium, 29, 61, 64
Chimneys
  for Aladdin lamps, 29–30, 32
  for kerosene wick lamps, 19–20
  hurricane, 10

Citronella, 7
Clay, for fuel reservoirs, 15–16, 24–25
Coleman Company, 39, 68–69, 71
Coleman fuel
  comparative fuel costs, 50
  See also Gasoline
Coleman pressure lanterns, 38–46
  incandescent mantles for, 62
Corning Glass, 73, 77, 79
Cressets, 12

D

Dietz lantern, 47–48
Duplex, 17, 18

E

Edison, Thomas, 63
Electricity, comparative fuel costs, 50
Evans, Robley D., 69–71

F

Falk Company, 64
Falk gaslight, 53
Fire-Brite, 78
Fireview stoves and fireplace inserts, 73–79
  sources for, 86–91
Flame spreader, for Aladdin lamps, 28
Franklin stoves, 76

Fuels
  for Aladdin lamps, 32
  for Coleman lanterns, 39, 43
  kerosene, 12, 14–15

G

Gallery, for Aladdin lamps, 28
Gaslights, liquid propane, 49–60
  sources for, 83–84
Gasoline, for Coleman lanterns, 39, 40–44
Gasoline pressure lamps, sources, 85
Glass
  for fuel reservoirs, 15, 24–25
  for stoves, 73–79

H

Heat shield, for LP gaslight, 54
High altitude, and Aladdin lamps, 30, 32
Humphrey Opalite, 53, 55, 56, 64
Humphrey Products, 64
Humphrey gaslight, incandescent mantle for, 62
Hurricane chimney, 10

I

Insect screen, for Aladdin lamps, 30

J

Johnson, Victor Samuel, 23–24

K

Kerosene, 12, 14–15
  comparative fuel costs, 50
  for Aladdin lamps, 32
  for Coleman lanterns, 39, 41–45
  source, 85
Kerosene pressure lamps, sources, 85
Kerosene pressure lanterns, 39
Kerosene wick lamps, 11–21
  sources for, 81–83

Kerosene wick lanterns, 37, 47–48
  sources for, 84–85

L

Lanterns, 37–48
  sources, 84–85
Lighting (starting)
  Aladdin lamps, 32, 35
  Coleman lanterns, 40, 44
  comparative fuel costs, 49–50
  LP gaslights, 59–60
Liquid propane (LP), 49–60
  for pressure lanterns, 39
  gaslights, 49–60
  vs. kerosene, 11–12

M

Mantles, incandescent, 23
  for Aladdin lamps, 28–29
  for Coleman lanterns, 40, 43
  for LP gaslights, 54, 57–59
  sources for, 85
  toxicity of, 61–72
Metal, for fuel reservoirs, 15, 24–25
Mother Earth News, 68–69

N

Neoceram, 73, 79
Nippon Electric Glass, 73, 77, 79
Nozzle, for LP gaslight, 54

P

Paraffin, 7
Porcelain, for fuel reservoirs, 24–25
Practicus, 23–24, 27
Preheater, for kerosene Coleman lanterns, 44
Pressure lamps
  indoor, 45–46
  sources for, 85
Pressure lanterns, 39–46

INDEX

Pump, for Coleman lanterns, 43–44
Pyrex, for stove glazing, 75, 79
Pyroceram, 73, 79

Q

Quick Light, 40

R

Reflectors
   for kerosene wick lamps, 20–21
   for LP gaslight, 54
Reservoirs
   for Aladdin lamps, 24–25
   for kerosene wick lamps, 15–16
Robax, 73, 77, 79

S

Safety tips and cautions
   for Aladdin lamps, 32, 34–36
   for candles, 8
   for Coleman lanterns, 40–42
   for kerosene wick lamps, 20–21
   for LP gaslights, 52–53
   for wood stoves, 78–79
Saveall, 9
Schott, 73, 77, 79
Snuffers, 10
Squeeze bulb, for Coleman kerosene lanterns, 44
Stoves, fireview, 73–79
   sources for, 86–91
Sulfur, 7, 14

Suppliers, addresses, 81–91

T

Tar, in kerosene, 14
Thorium, 29
   and radioactivity, 61, 63–72
TruLite, 14
Turnup wheel, 17

V

Valve cover, for LP gaslight, 54
Vermont Castings, 76
Vermont Health Department, 66–67
Vermont Occupational Safety and Health Agency, 67
Vestal, 15
Vycor, 79

W

Wagner, Walter, 68–72
Wall bracket, for LP gaslight, 54
Welsbach mantles. *See* Mantles, incandescent
Welsbach, Carl Auer von, 23, 63
Wick lamps, 11–21
Wicks
   for Aladdin lamps, 23, 28, 35–36
   for kerosene wick lamps, 17–19
Wilmar Associates, 64, 66
Wirth, Charles, 24